Understanding Immunology

Understanding Immunology

Second Edition

Peter Wood
University of Manchester

PEARSON
Prentice
Hall

Harlow, England • London • New York • Boston • San Francisco • Toronto • Sydney • Singapore • Hong Kong
Tokyo • Seoul • Taipei • New Delhi • Cape Town • Madrid • Mexico City • Amsterdam • Munich • Paris • Milan

Pearson Education Limited
Edinburgh Gate
Harlow
Essex CM20 2JE
England

and Associated Companies throughout the world

Visit us on the World Wide Web at:
www.pearsoned.co.uk

First published 2001
Second edition published 2006

© Pearson Education Limited 2001, 2006

The right of Peter Wood to be identified as author of this work has been asserted by him in accordance with the Copyright, Designs and Patents Act 1988.

ISBN 13: 978-0-13-196845-5
ISBN 10: 0-13-196845-9

British Library Cataloguing-in-Publication Data
A catalogue record for this book is available from the British Library

Library of Congress Cataloging-in-Publication Data
Wood, Peter (Peter John), 1952–
    Understanding immunology / Peter Wood.– 2nd ed.
        p. cm. –
    ISBN 0-13-196845-9 (paperback : alk. paper)
    1. Immunology. I. Title. II. Cell and molecular biology in action

    QR181.W735 2006
    616.07'9–dc22

                                        2005051506

10 9 8 7 6 5 4 3 2
11 10 09 08 07 06

Typeset in 9.5/12 Concorde BE Regular by 30.
Printed in Great Britain by Henry Ling Limited., at the Dorset Press, Dorchester, DT1 1H
The publisher's policy is to use paper manufactrured from sustainable forests.

For my wife, Jude
And in memory of Jack and Kirsten. *Elsket og Savnet*

Contents

Preface

This book is aimed at introducing immunology to students who have never studied the subject before. It is intended primarily for undergraduate students in the biological sciences and biomedical fields and medical students, although it is hoped it will appeal to other groups. No previous knowledge of immunology is assumed, but people reading this book will benefit from a basic knowledge of biochemistry and cell biology.

The impetus to write this book came from experience of teaching introductory immunology courses to undergraduate science, medical and dental students. Often there is a stage of bewilderment at the complexity of the immune system, which seems to utilise convoluted mechanisms to perform what are, on the face of it, simple tasks. Therefore one of the aims of this book is to try to explain why, in many situations, a more complicated-looking arrangement is actually much more efficient than an apparently simpler alternative.

The first part of the book is designed to take students step by step through the pathogenesis of infectious diseases and the molecules, cells and tissues of the immune system that provide protection against the wide variety of pathogens to which we are exposed. The latter part of the book deals more with the immune system operating in disease situations such as allergy, autoimmunity and transplantation. Inevitably, with an introductory book, some topics receive little or no attention. Therefore immunodeficiencies are referred to briefly and other topics such as xenotransplantation and mucosal tolerance are not covered.

The book is organised so that it is easiest to understand if the chapters are read sequentially. This has led to two basic rules in design. The first is that, wherever possible, new words and concepts are explained when they are first introduced, although inevitably reference is made to later chapters in some instances. Second, it is assumed in the later chapters that previous chapters have been read and an understanding of the previous concepts acquired. Where it is felt important to refer the reader to previous chapters this is done. Another concept of the book is the use of boxes. The material in the boxes is not essential for understanding of the rest of the text. The boxes are provided (i) to describe in more detail material covered in the text and (ii) to describe key observations or experiments that led to important advances in immunological knowledge (e.g. the discovery of the thymus), but they can be bypassed if time or inclination warrant this.

Chapter 1 begins with a description of the variety of pathogens, illustrating their diversity, and emphasises that no simple defence mechanism could deal with this variety of threats. The immediate response to infection is described in Chapter 2 and the requirement to distinguish between pathogens and host tissue introduced. Recognition by the innate immune system is covered. Chapter 3 describes the antibody molecule as part of the solution to having a better system of recognising pathogens than that provided by innate recognition and also introduces one of the cells of the specific immune system – the B lymphocyte. Other important types of lymphocyte, T cells, how they recognise antigen and the structure of the major histocompatibility complex (MHC) are described in Chapter 4. Chapter 5 describes the generation of a huge repertoire of antigen specificities from a limited number of genes. Chapter 6 describes the anatomy of the immune system and how it promotes the cellular interactions required in immune responses. The anatomical and cellular aspects of antibody production are expanded in Chapter 7. How antibody contributes to the elimination of pathogens or neutralises their effects is covered in Chapter 8. Chapter 9 is concerned with cellular immunity, which involves different types of immune responses than antibody production. One of the major successes of manipulating the immune system, vaccination, and its basis in immunological memory are described in Chapter 10. Chapter 11 deals with the production of B and T cells, explaining why lymphocytes must be continually produced. The remaining chapters are devoted to the immune system in disease. Chapters 12 and 13 cover autoimmunity and allergy respectively and suggest reasons for the severe increase in incidence of these diseases in developed countries, and Chapter 14 is about another immunological disease that has attracted attention – AIDS. The final chapter describes attempts, clinical and experimental, to manipulate the immune system in the fields of transplantation and tumour therapy. To reiterate, throughout the book attempts have been made wherever space has allowed to explain why the immune system is organised the way it is, so that its complexity can be understood.

Peter Wood

Acknowledgements

I would like to thank the many people who have contributed to the production of this second edition. First and foremost I thank Jude Wood for her extensive help with the figures and proofreading. I would also like to thank those who have provided assistance in the production process: Alex Seabrook, Louise Lakey, Bridget Allen and Pauline Gillett.

In addition the author and the publishers would like to thank the following:

We are grateful to the following for permission to reproduce copyright material:

Figure 15.4: The effect of *HLA-DR* matching on graft survival. Courtesy of Dr. Philip A. Dyer. Reprinted with permission; Plate 1: Elephantiasis. Courtesy of Richard Suswillo, Department of Infectious Diseases Epidemiology, Imperial College, School of Medicine, London, UK; Plate 2a, b and c: Cells of the phagocyte lineage, Plate 3a and b: Mast cells and basophils, Plate 4a and b: Lymphocyte and plasma cell, Plate 9: Histological section of part of the spleen (H&E × 20), Plate 10: Histological appearance of mucosal associated lymphoid tissue (MALT) and Plate 12: Histological appearance of thymus. All courtesy of Mike Mahon, John S. Dixon and Philip F. Harris, University of Manchester; Plate 5: Antibody structure. From Saphire, E.O., et al. (2001), 'Crystal structure of a neutralising human IgG against HIV-1: A template for vaccine design'. *Science*, Vol. 293, pp. 1155–59. Copyright © 2001 American Association for the Advancement of Science (AAAS). Reprinted with permission; Plate 6: Antibody-antigen interaction and Plate 8: Interaction between the T cell receptor (TcR) and antigen/class I MHC. Courtesy of Dr. Annemarie Honegger and Prof. Dr. Andreas Plueckthun, Biochemisches Institut Der Universitat Zurich, Winterthurerstrasse 190, CH-8057 Zurich, Switzerland.

Reprinted with permission; Plate 7: C1 I and C1 II MHC. From Stern, L.J. and Wiley, D.C., 'Antigenic peptide binding class I and class II histocompatibility proteins'. *Structure*, Vol. 2, pp. 245–51. Copyright © 1994 Elsevier. Reprinted with permission; Plate 11a and b: Langerhans cells and dendritic cells. Courtesy of Dr. Pedram Hamrah and Dr. Reza Dana, Schepens Eye Research Institute, Harvard Medical School, Boston, Massachusetts, USA.

In some instances we have been unable to trace the owners of copyright material, and we would appreciate any information that would enable us to do so.

CHAPTER ONE

The threat to the body:
the role and requirements of the immune system

Learning objectives

To be acquainted with the role of the immune system. To learn about the huge variety of threats posed to the body by infectious organisms. To appreciate that this requires a complex defence system.

Key topics

- The role of the immune system
- Types of pathogen
- Disease production by pathogens
  - ○ Infection
  - ○ Replication
  - ○ Spread
  - ○ Pathogenesis
- Barriers to infection

1.1 The role and complexity of the immune system

The immune system consists of proteins, cells and organs (Figure 1.1) that are concerned with **defence** of the individual, primarily against the threat of disease caused by **infectious organisms**. An infectious organism that causes disease is called a **pathogen** and the individual (person or animal) that is infected by a pathogen is called the **host**. Not all infectious organisms cause disease and some are actually beneficial, for example bacteria living in the gut help to digest certain foods. Infectious organisms that help the host are called **commensal organisms**. However, many viruses, bacteria, fungi, yeasts and parasites are pathogenic and we are constantly in danger of

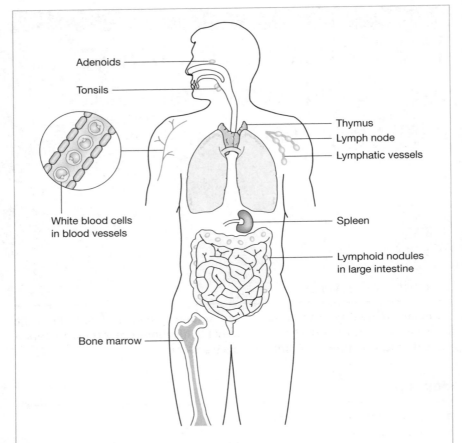

Figure 1.1 Some components of the immune system. The bone marrow and thymus are sites of production of lymphocytes, white blood cells involved in immune responses. Adenoids, tonsils, lymph nodes, spleen and lymphoid nodules are sites where immune responses are generated. Lymphatic vessels are similar to blood vessels and allow cells of the immune system to travel throughout the body.

infection and disease caused by them. Table 1.1 shows examples of infections caused by various pathogens.

The importance of the immune system in resistance to infection is demonstrated by the fact that, without treatment, babies born with severe defects of certain components of the immune system die in infancy of overwhelming infection, usually with organisms (e.g. *Haemophilus influenzae*, *Candida albicans* or cytomegalovirus) that cause little or no disease in individuals with a normal immune system. More recently, the AIDS (acquired immunodeficiency syndrome) epidemic has brought the consequences of having a poorly functioning immune system to the attention of the public at large. AIDS patients usually die from infection, such as with the yeasts *Pneumocystis carinii*, which causes pneumonia, or *Cryptococcus*

Table 1.1 Examples of infectious diseases caused by different families of pathogens

Organism	Disease
Viruses	
Hepadnavirus	Hepatitis B
Herpesvirus	Chickenpox
Poxvirus	Smallpox
Picornovirus	Polio, common cold
Myxovirus	Measles, mumps
Retrovirus	AIDS
Bacteria	
Streptococcus	Pneumonia
Clostridium	Tetanus, botulism
Neisseria	Gonorrhoea
Salmonella	Food poisoning
Vibrio	Cholera
Mycobacterium	Tuberculosis, leprosy
Fungi (yeasts and moulds)	
Trychophyton	Athlete's foot, jock itch
Candida	Thrush
Cryptococcus	Meningitis
Protozoan parasites	
Plasmodium	Malaria
Giardia	Giardiasis
Trypanosomes	Chagas' disease, sleeping sickness
Helminths	
Taenia	Tapeworm
Schistosoma	Schistosomiasis (flukes)
Ascaris	Roundworm
Onchocera	River blindness

neo-formans, a cause of meningitis. The immune system normally controls these infections with little or no damage to the host.

The immune system is primarily involved in defence against infectious organisms, but it also plays a role in various other diseases. Evidence is accumulating that the immune system can provide protection against some **tumours**. Exciting new developments also suggest that immune responses can be induced against tumours that normally do not provoke an immune response. These so-called tumour vaccines offer hope of additional weapons in the armoury against cancer. The immune system is also responsible for the rejection of **transplants**. In this instance the immune system is acting normally in trying to defend the body against a foreign invader, even though the 'invader' is beneficial. Rejection of transplants is a major cause of graft loss and much effort is being devoted to try to prevent transplant rejection.

Although a properly functioning immune system is essential for good health, like any other physiological system the immune system can malfunction and actually cause diseases. The most common form of disease caused by the immune system is **allergy**, for example hayfever and food allergy, which is caused by an inappropriate immune response against generally harmless material such as pollen or food. Another category of diseases caused by malfunction of the immune system are the **autoimmune diseases**, including rheumatoid arthritis and some types of diabetes. Autoimmune diseases occur when the immune system attacks the body's own tissue. Although these diseases caused by the immune system can be threatening and sometimes fatal, on balance a properly functioning immune system is essential for life in the world we inhabit, which is full of disease-causing organisms, some of which, as mentioned above, are invariably fatal if the system does not operate properly.

1.1.1 Why is immunology so complicated?

There are thousands of components to the immune system, and during the course of learning about some of these it can appear that the immune system is far more complicated than necessary for achieving what is, on the surface, the simple task of eliminating an infectious organism. There are a number of reasons why the immune system is complex. The first of these is the desirability of eliminating pathogens without causing damage to the host. Getting rid of a pathogen is theoretically easy. If you had an infection in your liver you could produce a nasty toxin that would kill the pathogen; unfortunately it would also destroy your liver. Killing pathogens is not difficult, but getting rid of pathogens without damaging the host is much more complicated. Imagine if a city in your country was infiltrated by soldiers from another country; you could get rid of the foreign invaders by dropping a nuclear bomb on the city but this would also kill a lot of your compatriots. To go into the city to eliminate or capture the foreign soldiers without causing harm to your compatriots is much more difficult. This analogy raises a major issue concerning the immune system – that of **recognition**. To eliminate foreign soldiers without killing your own requires that you can tell the two apart. In the same way the immune system must be able to distinguish between pathogens and host cells so that it can direct its destructive powers towards the pathogens. Many of the specialised features of the immune system are involved in recognition of foreign pathogens.

An additional problem facing the immune system is that pathogens come in all shapes and sizes, with different lifestyles and different ways of causing disease. To understand fully the complexity that the immune system must deal with, it is necessary to have some understanding of infectious organisms and the ways in which they cause disease. The rest of this chapter describes how pathogens differ, so that it is possible to get an appreciation of the problems faced by the immune system.

1.2 Types of pathogen and how they differ

The types of pathogen that can cause disease include many groups of single-celled microorganisms and larger multicellular parasites. Viruses, bacteria, some yeasts, and protozoan parasites are examples of single-celled pathogens. Fungi and helminths (parasitic worms) are the major

Table 1.2 Size and lifestyle of pathogens

Organism	Size	Habitat	Mode of multiplication	Multiplication rate (doubling time)
Viruses Poliovirus	20–400 nm	Intracellular: pharynx, intestine, nervous system	Intracellular synthesis of viral components	<1 hour
Poxvirus		Intracellular: upper respiratory tract, lymph nodes, skin	Intracellular synthesis of viral components	<1 hour
Bacteria Streptococcus pyogenes	1–5 μm	Extracellular: pharynx	Cell fission	3 hours
Mycobacterium leprae		Intracellular: macrophages, endothelial cells, Schwann cells	Cell fission	2 weeks
Fungi Candida albicans	2–20 μm	Extracellular: mucosal surfaces	Asexual budding	Hours
Histoplasma capsulatum		Intracellular: macrophages	Asexual budding	Hours
Protozoan parasites Trypanosomes	1–50 mm	Extracellular: bloodstream	Binary fission	6.5 hours
Plasmodium		Intracellular: red blood cells, hepatocytes	Asexually in hepatocytes (cell fission)	8 hours
Metazoan parasites (worms) Ascaris lumbricoides	3 mm to 7 m	Intestine	Lays eggs	200 000 eggs/day
Taenia solium (tapeworm)		Gut	Releases body segments containing eggs	800 000 eggs/day

multi-cellular pathogens (Table 1.1). These pathogens come from very different parts of the biological kingdom and vary considerably in many aspects. Pathogens differ enormously in their size. They also have very different lifestyles and cause disease in a variety of ways (Table 1.2).

1.2.1 Size of pathogens

One feature of the range of pathogenic organisms listed in Table 1.2 is the enormous variation in **size**. Viruses are the smallest infectious organisms, being 20–400 nm in size. At the other end of the scale some parasitic worms, such as the tapeworm, can be up to 7 m (20 ft) in length. This represents a difference in scale of a factor of 10^9. To put that into some sort of perspective, if a virus were the size of a tennis ball, a fully developed tapeworm would reach from London to Los Angeles. It does not stretch the imagination too far to appreciate that the problems posed to the immune system by these two organisms would require very different solutions.

1.2.2 Stages of disease production by pathogens

Size is not the only way in which infectious organisms vary. They also vary enormously with respect to how they enter and live within the body and actually cause disease. Infection and disease production by pathogenic organisms can be divided into four stages:

1. Invasion.
2. Multiplication.
3. Spread.
4. Production of disease (pathogenesis).

Although infection usually involves all of these steps, there are many exceptions in terms of both the steps involved and their order. Some pathogens do not spread significantly or even technically gain entry to the body. Organisms may replicate locally before spreading or may spread through the body before beginning significant replication. Pathogens show considerable variation at each of these stages of infection, as will be described below.

1.3 Disease production by pathogens

The first stage of disease production by pathogens is infection, or entry of the pathogen into the body. Nearly all pathogens must gain entry into the body before they can begin to replicate or spread. A few pathogens can exist on the skin (e.g. viruses causing warts) or in the gut (e.g. bacteria causing cholera) without technically entering the body. However, infection is not made easy for pathogens because the body has many physical and chemical barriers to try to prevent pathogens entering the body.

1.3.1 Barriers to infection

The body has many physical, chemical and biochemical barriers that make it much more difficult for pathogens to gain entry into the body (Figure 1.2).

Figure 1.2 Physical and chemical defence mechanisms.

The physical barriers to infection are as follows:

- **Skin and mucosa.** Intact skin and mucosa provide a physical barrier to prevent entry of organisms.

- **Cilia.** The respiratory tract is lined with little hair-like structures that beat in such a way as to propel particles towards the throat, where they can be expelled by coughing or swallowing and excretion.

- **Mucus.** Mucus is secreted by epithelial cells of the gut, respiratory tract and genito-urinary (GU) tract. It has the unusual properties of being sticky and slimy at the same time and is able to entrap microorganisms so they can be expelled. In the respiratory tract, cilia and mucus combine to provide an effective way of trapping and eliminating microbes.

The **chemical and biochemical** defences are as follows:

- **Acids.** Hydrochloric acid secreted by the stomach is lethal to many (though not all) bacteria. Commensal bacteria in the vagina produce lactic and proprionic acid resulting in a low pH, which is inhibitory to the division of many bacteria.

- **Fatty acids.** Sebaceous glands in the skin produce fatty acids that have antimicrobial properties.

- **Lysozyme.** This is present in sweat, tears and many other secretions. It breaks down peptidoglycans in bacterial cell walls, thus damaging and killing the bacteria.

- **Defensins.** Defensins are antimicrobial peptides that are found in the secretions of mucosa and skin.

- **Cathelicidins.** These antibacterial peptides were originally discovered as insect defence peptides. Other members of the cathelicidin family have been found in mucosal secretions.

- **Collectins.** Collectins are proteins that can bind sugars on microbial surfaces and promote the elimination of microbes. Proteins that bind sugars are known as lectins; because collectins bind sugars in a calcium-dependent manner, they are known as C-type lectins. The A and D lung surfactants are collectins that provide protection at the lung surface; other collectins, such as mannose-binding protein, are found in serum.

The physical and chemical barriers are very effective at preventing pathogens from entering the body and they exclude more than 99.9% of the infectious organisms we are exposed to. However, organisms do infect the body. This can occur in a number of ways.

1.3.2 Invasion – entry of pathogens into the body

Routes by which infectious organisms gain entry into the body include the skin, respiratory tract, gastro-intestinal (GI) tract and GU tract. There are fundamentally two ways in which infectious agents cross the physical and chemical barriers: either they are able to penetrate the intact barriers at one or more anatomical sites, or the physical barriers are damaged and breached, allowing entry of the organism (Figure 1.3).

Penetration of intact skin or mucosa

- **Skin.** Few organisms are able to penetrate intact skin. However, some parasites (e.g. hookworm) or their larvae (e.g. schistosoma) can do this. Other agents, such as wart viruses, set up infection in the skin and do not enter further into the body.

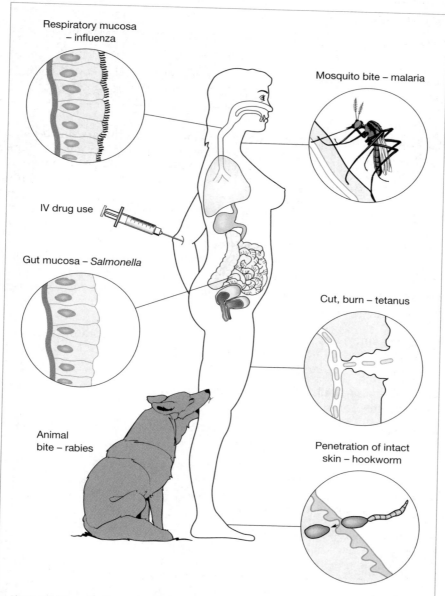

Respiratory mucosa
– influenza

Mosquito bite – malaria

IV drug use

Gut mucosa – *Salmonella*

Cut, burn – tetanus

Animal
bite – rabies

Penetration of intact
skin – hookworm

Figure 1.3 Entry of pathogens into the body. Insect bites, cuts, burns and animal bites breach the skin barrier, allowing entry of pathogens. Some parasites can penetrate intact skin while many pathogens penetrate intact mucosa of the respiratory, intestinal and genito-urinary tracts.

Table 1.3 Mucosal sites of entry for pathogens

Pathogen	Disease	Mucosal site of entry
Rhinovirus	Common cold	Nasal epithelium
Influenza virus	Influenza	Upper respiratory tract
Bordatella pertussis	Whooping cough	Lower respiratory tract
Salmonella spp.	Food poisoning	Small intestine
Rotavirus	Diarrhoea	Small intestine
Escherichia coli (some strains)	Urinary tract infection	Bladder, ureter
Neisseria gonorrhea	Gonorrhoea	Vagina, urethra

- **Mucosa.** Mucosa, being softer and damper than skin, are much more frequent sites of entry and all intact mucosa can be penetrated by some organisms. Examples are shown in Table 1.3. Pathogens can cross epithelia by passing through epithelial cells, as in the case of the meningococcus (a bacterium causing meningitis), or by passing between the epithelial cells, seen with *Haemophilus influenzae*.

Penetration of damaged skin or mucosa

There are many ways in which skin or mucosa can be damaged, allowing entry of infectious organisms that could not cross intact skin or mucosa. Damage to skin is a particularly important route of infection and can occur in a number of ways:

- **Burns.** Burns, especially severe ones, pose a major risk for infection, particularly with *Staphylococcus, Streptococcus, Pseudomonas* and *Clostridium tetanus*.

- **Cuts and wounds.** These can allow entry of similar organisms to those seen after burns.

- **Insect bites.** Numerous infections are transmitted via insect bites. These include malaria, typhus and plague.

- **Animal bites.** Animal bites can provide direct transmission of infection, such as in rabies. Because they cause significant damage to the skin, bites can allow the entry of the same environmental pathogens as burns, cuts and wounds (see above).

- **Human behaviour.** Various aspects of uniquely human behaviour can result in the skin being penetrated. Sharing of syringes by intravenous (IV) drug users exposes them to risk of hepatitis and human immuno-deficiency virus (HIV). A number of viral infections (hepatitis, HIV) have been transmitted by blood transfusion and blood products (e.g. factor VIII for haemophiliacs) before appropriate screening procedures were developed. Transplantation has also resulted in transmission of infection before the introduction of appropriate donor screening.

Damage to mucosa may not increase the likelihood of infection to the same extent as damage to the skin. However, physical or chemical damage may allow entry of some organisms (e.g. smoking increases the risk of respiratory bacterial infections). Furthermore, infection of the mucosa with a virus may cause damage and facilitate the entry of bacterial pathogens.

1.3.3 Multiplication of pathogens

Most initial infections are local, i.e. the infectious agent gains entry to the body at a single site, e.g. via an insect bite or infection of a particular mucosal surface. The next stages of infection involve multiplication and spread of the pathogen. These can be considered part of the lifestyle of the pathogen, and infectious organisms vary enormously in lifestyle.

Multiplication of pathogens provides variety at three levels: the **mode** of multiplication, the **site** of replication and the **rate** of multiplication.

Mode of multiplication

Different pathogens multiply in very different ways (Figure 1.4). Many single-celled organisms, including bacteria, yeasts and protozoan parasites, divide by simple cell division. Viruses, however, have a completely different mode of multiplication called replication. Following infection of a cell, viral particles disassemble and, under direction of viral nucleic acid (DNA or RNA), new viral proteins and genetic material are synthesised. Eventually new viral particles are assembled and leave the cell. This can occur by the cell bursting open and releasing viral particles to infect other cells, resulting in cell lysis and death of the cell. Alternatively the cell can shed viral particles in a more gradual manner, a process known as budding, which does not result in the death of the cell. Finally many parasitic worms do not multiply directly but lay eggs, which provide additional sources of infection for other organisms.

Site of replication

Pathogens can live and multiply inside host cells or outside the cells. Many bacteria, yeasts and parasites multiply extracellularly. Viruses by their nature have to replicate intracellularly because they lack enzymes and other cofactors necessary for synthesising viral proteins. Many bacteria and protozoan parasites also replicate intracellularly. Some organisms can live in either an intracellular or an extracellular environment (e.g. *Mycobacterium tuberculosis, Neisseria gonorrhoeae*). Parasites (e.g. trypanosomes) have the most complicated life cycles, which can often involve both an intracellular and extracellular stage.

The site in which pathogens live and multiply poses different problems for the immune system. The most important of these is whether the pathogen has an intracellular stage, because during this stage the organism may be partially hidden from the immune system. However, as we shall see in Chapter 7, the immune system has even evolved ways of detecting whether infected host cells are harbouring hidden pathogens.

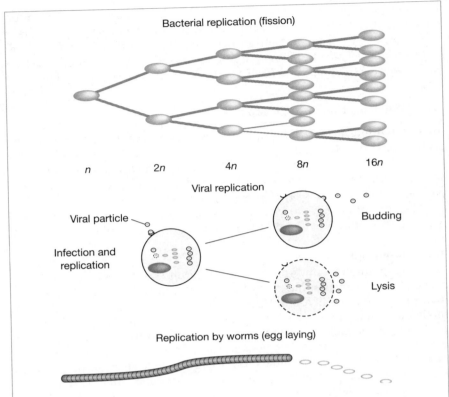

Figure 1.4 Multiplication of pathogens. Many bacteria divide by simple cell fission. Viruses must infect host cells to replicate. Parasitic worms (helminths) often lay eggs, which are transmitted to new hosts before developing into worms.

Rates of multiplication
The time taken for pathogens to reproduce themselves varies enormously. Some bacteria under optimal conditions *in vitro* can divide every 20 minutes. At this rate of division a single bacterium would produce over 10^{21} progeny in a day! Obviously this rate of replication is unsustainable for long, even under optimal *in vitro* conditions, and it is debatable whether it is ever reached *in vivo*. Viral replication can result in hundreds or thousands of progeny being produced from a single virion in hours. Other pathogens have low rates of replication. Not all bacteria have the capacity to divide rapidly and some, such as the mycobacteria, the causes of tuberculosis and leprosy, have a doubling time of many days. Some parasitic worms never replicate within the host, although they may lay eggs, thereby increasing the number of organisms that can infect other hosts. However, again the rate of egg laying can vary enormously: *Schistosoma mansoni*, the cause of the disease schistosomiasis, lays only 200 eggs a day while *Ascaris lumbricoides*, a roundworm, may lay over 200 000.

The way in which organisms spread through the body is influenced to some extent by whether they live intracellularly, extracellularly or both. Organisms that live extracellularly are able to spread via body fluids such as blood. However, even organisms that replicate intracellularly may be able to leave the cell and spread via an extracellular route. Organisms can spread in the following ways:

- **Cell to cell contact.** Many organisms, especially viruses, spread directly from cell to cell with essentially no extracellular component to their lifestyle. These pathogens tend to cause localised infections such as seen in influenza, where only the respiratory tract is infected. However, localised infections can still cause widespread symptoms, so that 'flu causes headache, fever and muscle-ache.

- **Via blood and lymphatic vessels.** The commonest, and fastest, way in which pathogens can spread through the body is via the bloodstream. Since all organs and tissues require a blood supply, microorganisms in the blood have the potential to spread to all sites. However, individual pathogens show a preference to localise in particular organs or tissues that may differ from pathogen to pathogen (see Table 1.2).

  The lymphatic vessels form a circulatory system that parallels that of the blood (Figure 1.1; see also Chapter 6). There are important differences between the two systems, however. The circulation of the lymphatic fluid is maintained not by the heart but by the movement of the muscles surrounding the lymphatic vessels; thus lymphatic fluid flows at a much more sluggish rate than blood. Moreover, tissue fluid can drain directly into lymphatic vessels. Organisms can easily enter lymphatic vessels draining the site of infection, where they will be conveyed to the local lymph nodes.

- **Spread via body cavities.** Microorganisms that have infected one organ in a body cavity such as the peritoneum may occasionally spread via the cavity to other organs located within it.

- **Spread via nervous system.** This is a particularly important route of spread for certain viruses. Viruses can spread via peripheral nerves to the central nervous system (CNS) or vice versa. In some instances this route of spread allows the virus to become more widespread within the nervous system where it resides and causes disease (e.g. herpes simplex virus). In other cases the virus travels via nerves to infect other organs. The rabies virus infects the salivary glands in this way, enabling the virus present in the saliva to be transmitted via a bite.

1.3.5 Pathogenesis

The final stage of the disease process (although it may not be the final stage of the infection) is the actual production of disease. Many microorganisms live in or on the body without causing disease. These organisms are called **commensal** organisms and may be beneficial to the host: the production of lactic and proprionic acids by lactobacilli in the vagina inhibits the growth of many other bacteria and many commensal organisms compete with pathogens for 'living space' in the gut. Pathogens differ in that they cause disease by one or more mechanisms (Figure 1.5). These include the following:

- **Secretion of toxins.** Many organisms, especially bacteria, secrete toxins that either directly or indirectly account for most of the pathology caused by the organism. These include the powerful neurotoxins secreted by the *Clostridium* family of bacteria responsible for tetanus or botulism food poisoning, toxins of the bacteria *Shigella dysenteriae* and *Vibrio cholerae* that cause dysentery and cholera respectively, and toxins secreted by *Streptococcus pyogenes*, which can cause scarlet fever (Box 1.1). Some protozoa and fungi also secrete exotoxins.

- **Endotoxins.** Endotoxins, rather than being secreted, are components of the cell wall of pathogens. They are particularly prevalent in Gram-negative bacteria (e.g. *Salmonella*) but are also found in other bacteria, some yeasts and protozoa (Box 1.1). Unlike exotoxins, which have

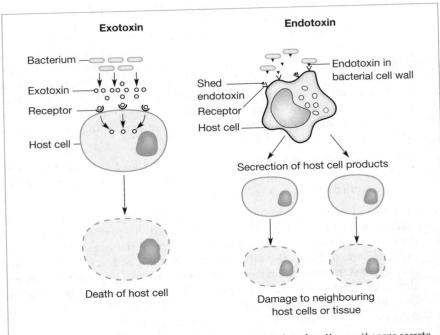

Figure 1.5 Damage caused by microbial exotoxins and endotoxins. Many pathogens secrete exotoxins, which bind to molecules on host cells, enter the host cell and kill it. Endotoxins are components of the cell wall of some pathogens. Endotoxins in the cell wall, or shed endotoxin, bind to receptors on certain host cells (such as macrophages) and stimulate the host cells to secrete products that damage neighbouring host cells and tissue.

Exotoxins

Exotoxins are secreted products, usually of bacteria but sometimes protozoa and fungi. They can act in a number of ways:

- **Inhibition of protein synthesis.** *Corynebacterium diphtheriae*, the cause of diphtheria, produces a toxin that causes ADP-ribosylation of elongation factor-2, thereby stopping protein synthesis. It is extremely potent and one molecule of toxin is capable of killing a cell. *Escherichia coli, Vibrio cholerae* and *Bordatella pertussis* (the cause of whooping cough) also produce toxins that cause ADP-ribosylation of proteins.

  Toxins from *Shigella dysenteriae* and *E. coli* strain O157:H7 (a cause of dangerous food poisoning) inhibit protein synthesis by removing adenine from 28s rRNA.

- **Increase in cAMP.** A number of bacteria produce toxins that raise cAMP levels. These include *V. cholera, Bacillus anthracis, B. pertussis* and some strains of *E. coli*. The consequence of increased cAMP levels is alterations in ion transport and hence fluid movement, often resulting in severe oedema.

- **Neurotoxins.** Members of the *Clostridium* family produce particularly potent neurotoxins. *Clostridium tetani* produces a toxin that prevents the release of glycine, an inhibitory neurotransmitter. This results in overactivity and muscle spasm including the typical lockjaw. *Clostridium botulinum* produces a neurotoxin that stimulates release of acetyl choline, leading to paralysis. It is one of the most potent toxins known and it is estimated that less than 1 μg can kill a person.

- **Enzymes that disrupt cell walls.** *Clostridium perfringens*, a cause of gas gangrene, produces a toxin called α-toxin, which is a phospholipase that hydrolyses lecithin in the cell membrane, resulting in cell death.

- **Superantigens.** Some bacteria, particularly *Staphylococcus* and *Streptococcus*, produce toxins that cause excessive stimulation of the immune system (specifically of T lymphocytes; see Chapter 6). This leads to the production of factors by the immune system that cause the symptoms of shock. Toxic shock syndrome and food poisoning are two consequences of these toxins.

direct, very specific toxic effects, endotoxins act by causing cells of the host to produce factors that cause fever, a fall in blood pressure and other symptoms.

- **Direct killing of host cells.** Some intracellular dwelling pathogens replicate within cells and leave the cells (usually by budding from the cell surface) with relatively little damage to the cell. This results in the continuous production of infectious particles by an infected cell. Other pathogens replicate within the cell and kill the cell, which bursts open (a process called cell lysis), thereby releasing many infectious particles (see Section 1.3.3). Many viruses and protozoa lyse host cells in this way; if this lysis is extensive enough, it will result in disease.

- **Physical blockage.** Larger pathogens may cause pathology purely by their physical presence. Probably the most dramatic example of this is elephantiasis caused by the filarial worms. By blocking lymphatic drainage these organisms can cause massive swelling of the breasts, testes and legs (Plate 1).

1.4 Conclusion

From the above description of the variety of pathogens and the way they live and cause disease, it can be appreciated that the immune system is faced with an enormous variety of problems when trying to protect the body from disease caused by all the different types of pathogens. Box 1.2 summarises the lifestyles of two pathogenic organisms. There is one

---

BOX 1.2: EXAMPLES OF THE LIFESTYLE OF PATHOGENIC ORGANISMS

Measles

The measles virus enters the body through the respiratory tract. It then travels to local lymph nodes and lymphoid tissue located in the mucosa. After a few days the virus spreads to other lymphoid tissue, including the spleen, where it begins to replicate. After a week or so, large quantities of the virus spread via the blood-stream to epithelial sites throughout the body. The presence of large amounts of virus at these sites gives rise to the various symptoms seen in measles. Virus in the respiratory tract causes runny nose and coughing. There is inflammation of the conjunctiva, and the presence of the virus in the skin causes the characteristic rash seen in measles.

Typhoid

If the *Salmonella typhi* bacterium is ingested and the dose is big enough, some bacteria will survive the acid environment of the stomach and enter the intestine. Bacteria penetrate the gut mucosa through specialised lymphoid structures known as Peyer's patches (see Chapter 6) and spread to the intestinal lymph nodes, where they proliferate in macrophages. Eventually the organisms reach the bloodstream, where they spread mainly to the liver, bone marrow and spleen, where they continue to multiply. This results in a further large increase in bacterial numbers and subsequent spread of the organism to other tissues such as the kidney and the gall bladder via blood or the biliary tract. The bacteria can also spread to the brain, heart and skin. The bacteria then invade the intestinal tract in much larger numbers than seen with the original infection and cause inflammatory lesions in the Peyer's patches, which may result in ulceration of the intestinal wall. The presence of the bacteria in other sites may cause meningitis, osteomyelitis, endocarditis and rashes.

additional factor that further challenges the immune system and increases the complexity of the immune responses required. Pathogens have co-evolved with the immune system and have developed survival strategies to counter attempts to eliminate them. It is obviously in the pathogen's best interests to survive in the host, and natural selection occurs so that pathogens with an improved ability to survive and multiply within hosts will have a selective advantage and become more common. The evolution of some pathogens seems to have been strongly influenced by the need to evade the immune response; for instance, cytomegalovirus (a cause of pneumonia) has devoted 30% of its genome to subverting the immune response against it. The immune system has accordingly had to evolve an equally complex variety of mechanisms to deal with the wide range of threats posed by different pathogens.

1.5 Summary

- The body is continually exposed to infectious organisms that have the potential to cause disease (pathogens). Most pathogens are prevented from entering the body by a combination of physical, chemical and biochemical defence mechanisms. However, some pathogens can breach the barriers and in some cases the barriers are breached by injury or other causes.

- Pathogens vary enormously in terms of size, ways in which they enter the body, how they multiply, whether they replicate intra- or extracellularly, replication rates, mechanisms by which they spread through the body and ways in which they actually cause disease.

- The variety of pathogenic lifestyles means that the immune system must have an equally varied repertoire of mechanisms for dealing with the diversity of threats.

CHAPTER TWO

The immediate response to infection: innate immunity and the inflammatory response

Learning objectives

To be aware of the stages of the immune response to infection. To understand the concept and molecular aspects of recognition of foreign organisms by the innate immune system. To learn about the inflammatory and acute phase responses and how leukocytes move around the body.

Key topics

- The response to infection
- Recognition of pathogens by the innate immune system
- Cytokines
- Cell migration
- The inflammatory response
  - Cellular components
  - Activation of the complement, kinin and clotting systems
- The acute phase response
  - Effect on the brain
  - Effect on the liver
  - Phagocytosis and opsonins
- Natural killer cells and interferons
  - Phagocytosis and opsonins
  - Natural killer cells

2.1 The response to infection

Chapter 1 covered pathogenic organisms and the stages of infection leading to disease, illustrating the tremendous variety of organisms and disease processes to which the body is exposed. Now we turn to the problem of how the body attempts to counter the threats posed by this enormous number of different infectious organisms.

The response to infection can be divided into five stages:

1. **Awareness** of infection. Obviously the body cannot begin to mount a defensive response against a pathogen until it is aware of the presence of the pathogen.
2. The **immediate** response to infection. This involves the activity of cells and other factors that are present at the time of infection but may require their recruitment to the site of infection and activation once there.
3. The **delayed** response to infection. If the immediate response to infection is not effective in eliminating the pathogen, the next stage is to generate new cells and factors to deal with the infection.
4. **Destruction** or **elimination** of the pathogen or **neutralisation** of the threat posed by pathogens. The optimal way of dealing with pathogens is to kill them or eliminate them from the body. However, in some cases where pathogens are producing a powerful toxin it may be more beneficial to neutralise the toxin first before attempting to destroy or eliminate the pathogen.
5. Provision of **immunity** so that you do not get ill if you are infected again with the same pathogen.

2.2 The immediate response to infection – the innate immune system

The term 'innate immune system' is used to describe pre-existing defence mechanisms that are designed to prevent infection by pathogens or to mount an immediate defence against the infectious agent. The physical, chemical and biochemical barriers to infection described in Chapter 1 are part of the innate immune system. They are called 'innate' or 'natural' because they are present before infection, although the amount of some components may increase following infection.

2.2.1 Cells of the innate immune system

There are also many cells and proteins found throughout the body that are part of the innate immune system and are involved in defence against pathogens. Collectively these cells and proteins perform two important functions: they are able to recognise the presence of a foreign entity, for example a bacterial infection, and they provide an immediate cellular response to the presence of an infectious agent. The nature of this cellular response differs according to the nature of the cell and the way in which it is stimulated.

One important cell type of the innate immune system is the **macrophage**, which is a bone marrow-derived cell type found in most tissues

(Plate 2). Some macrophages reside for a long time in the same tissue and are therefore called fixed macrophages. Other macrophages live up to their name of wandering macrophages by moving through tissue. Macrophages are related to a type of white blood cell called a monocyte (Plate 2), which is also derived from bone marrow. Monocytes can leave the bloodstream and enter tissue, where they differentiate into macrophages. Because of their relationship, macrophages and monocytes are known as cells of the monocyte/macrophage lineage.

Another cell type involved in protection is the neutrophil (Plate 2), which is also called a poymorphonuclear leucocyte (PMN). This is another type of blood cell that can leave the bloodstream and enter damaged or infected tissue. Other cell types found in tissues include mast cells and dendritic cells, which have special functions described later (see Chapter 13 for more detail on mast cells and Chapter 8 for information on dendritic cells).

2.2.2 Recognition of pathogens by cells of the innate immune system

When we say that cells of the innate immune system can 'recognise' pathogens, what do we actually mean? By recognition we mean that molecules, or receptors, on cells of the innate immune system bind to other molecules that are present on pathogens but not present on our own cells. In this way the immune system can distinguish foreign objects and respond to a foreign pathogen but not respond against our own tissue or cells. It is a crucial feature of the immune system that it can target foreign objects. There are many receptors present on cells of the innate immune system and some of them are shown in Table 2.1.

An important family of receptors that are able to recognise a wide variety of pathogen-associated molecules has been identified. These are the Toll-like receptors (TLRs), which were named because of their structural relationship to Toll, a protein involved in development of *Drosophila*, the fruit fly. There are about 10 TLRs that have been identified in man and similar numbers in other species. The TLRs recognise a variety of microbial products such as double-stranded RNA (found only in viral infection), lipopolysaccharide (LPS) from bacterial cell walls, bacterial lipoproteins, unmethylated DNA (indicating it is of bacterial origin) and flagellin, a component of bacterial flagella. Different TLRs recognise different microbial products, as shown in Table 2.2.

2.2.3 The cellular response to recognition of microbial products

It can be seen that many cell types of the innate immune system have receptors on their surface that can recognise microbial products and therefore detect that we have been infected. Obviously recognition of infection is useful only if it results in a response to the infection that will ideally eliminate the pathogenic organism or at the very least limit the replication and spread of the pathogen. There are a number of ways in which different cells can respond to recognition of pathogens, and one of these is phagocytosis.

Table 2.1 Phagocyte receptors

Receptor	Cellular distribution (phagocytes)	Pathogen molecules recognised	Pathogen distribution
Mannose receptor	Macrophages, neutrophils	Mannose-containing carbohydrates (polysaccharides)	Many bacteria
Scavenger receptor	Macrophages	Sialic acid	Bacteria and yeast
CD14	Macrophages, neutrophils	Lipopolysaccharide (LPS) component of bacterial cell walls	Gram-negative bacteria
Complement receptors CR3 and CR4	Macrophages	LPS, lipophosphoglycans	Bacteria and yeast

Table 2.2 Recognition by Toll-like receptors

Toll-like receptor	Ligands
TLR1	Bacterial lipopeptide
TLR2	Peptidoglycans, lipopeptides, zymosan (from yeast)
TLR3	Double-stranded RNA (viral)
TLR4	LPS (Gram-negative bacterial cell wall)
TLR5	Flagellin
TLR6	Lipopeptide, zymosan
TLR7	Single-stranded RNA
TLR8	Single-stranded RNA
TLR9	Unmethylated DNA (bacterial)
TLR10	? Unidentified molecules on some bacteria

2.2.4 The cellular response to infection – phagocytosis

Phagocytosis is the ingestion and destruction of microbes by cells called phagocytes. The two main types of phagocytes are the macrophages and neutrophils described above. The way in which macrophages and neutrophils phagocytose particles is essentially the same and can be divided into four stages (Figure 2.1):

1. **Attachment** of the phagocyte to the particle being phagocytosed, which may be a pathogen, a dead or damaged host cell or a piece of tissue.

2. **Ingestion.** By extending membrane protrusions called pseudopodia around the particle, the phagocyte is able to engulf the particle, which is taken into the cell in a phagocytic vacuole.

3. **Killing.** If the ingested particle is a live cell of a pathogen (e.g. a bacterium) the phagocyte will normally kill the cell by one of a number of mechanisms (described in more detail in Chapter 8).

4. **Degradation.** The phagocytosed particle, whether it is a dead cell or a piece of tissue, is broken down by enzymes in the phagocytic vacuole.

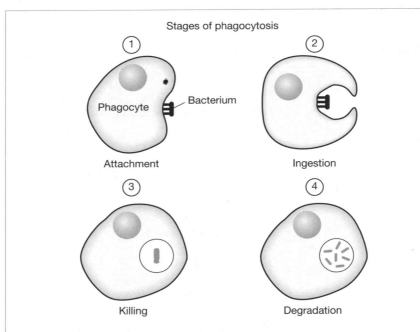

Figure 2.1 Phagocytosis. Phagocytes can take up and remove bacteria and dead host cells or tissue debris. The figure shows phagocytosis of a bacterium: ① The phagocyte binds to the bacterium. ② The phagocyte extends projections around the bacterium and engulfs it in a phagocytic vacuole. ③ The phagocyte kills the engulfed bacterium. ④ The bacterium is degraded by proteolytic enzymes.

Although the basic process of phagocytosis is similar in neutrophils and macrophages, there is an important difference. While neutrophils are only able to phagocytose small organisms such as bacteria and viruses, macrophages are able to phagocytose larger particles such as dead cells and tissue debris in addition to microorganisms. Therefore macrophages are involved in eliminating pathogens from tissues and also in cleaning up damaged tissue by removing dead or damaged host cells. Macrophages are able to distinguish between healthy host cells and dead/damaged cells because the receptors they have for recognising sugars on microbes also recognise sugars that are exposed by dead or damaged host cells (see Figure 2.2).

2.2.5 The cellular response to infection – production of new factors

When cells of the innate immune system encounter pathogenic products they can respond in ways other than phagocytosis. Cells can be stimulated to synthesise and/or secrete an enormous variety of new products. Some of these products may be directly involved in killing pathogens. Other products are involved indirectly in recruiting other cell types to try and eliminate the pathogen. An important group of proteins that can be secreted in response to pathogenic stimuli are known as cytokines.

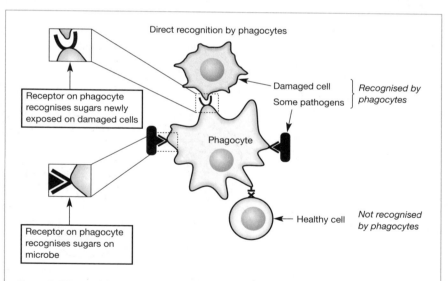

Figure 2.2 Recognition by phagocytes. Phagocytes must distinguish microbes and dead host cells from healthy host cells so that healthy host cells are not phagocytosed. Phagocytes have receptors on their surface that recognise sugars present on microbes or sugars that are newly expressed on dead or damaged host cells. These sugars are not present on healthy host cells and therefore the host cells are not phagocytosed.

2.3 Cytokines

The term cytokine covers a large number of smallish proteins (usually less than 20 kDa) that serve a hormone-like function in enabling cells to communicate with each other. Most people are familiar with hormones such as insulin and growth hormone, which are produced in one organ or tissue and travel through the bloodstream to other organs where they bind to receptors on the cells of that organ and stimulate a particular response. Hormones that are produced in one organ and act on a distant tissue are said to be acting in an **endocrine** manner (Figure 2.3). Cytokines do not usually act in an endocrine manner; rather, they act locally. They are produced by cells in a particular tissue and act on 'cells' in that tissue. Cytokines therefore act in a **paracrine** or **autocrine** manner (Figure 2.3). Paracrine action means that the cytokine binds to receptors on cells close to those producing the cytokine; by 'close' we are probably talking about a microenvironment of microns to 1 mm. Autocrine means that the cytokine actually binds to receptors on the cell that produced the cytokine. Thus the role of cytokines is to enable cells to communicate with each other in a local environment. A few cytokines can also act in an endocrine manner.

There are many cytokines and they can be divided into families (Table 2.3). The main families of cytokines are the interleukins, colony-stimulating

Figure 2.3 Action of hormones. Endocrine: the hormone is secreted at one site of the body and travels through the bloodstream. The hormone will bind to receptors on cells at a distant site (blue cells) and cause a response in those cells. **Paracrine:** hormones produced by cells in a tissue bind to receptors on other cells in the immediate vicinity (blue cells). Cells in other parts of the same tissue are not affected by the hormone (white cells). **Autocrine:** the secreted hormone binds to receptors on the cell that produced the hormone (blue cell).

Table 2.3 Cytokine families

Family	Members	Comments
Interleukin (IL)	IL-1 to IL-32	Different IL have different functions and are secreted by different cells.
Interferon (IFN)	IFNα	Leucocyte IFN. Inhibits viral replication.
	IFNβ	Fibroblast IFN. Inhibits viral replication.
	IFNγ	Secreted by lymphocytes. Many immunoregulatory functions.
Tumour necrosis factor (TNF)	TNFα	Secreted by monocytes and other cells. Factor activates macrophages and endothelium.
	TNFβ	Secreted by T cells. Similar activity to TNFα.
Colony-stimulating factor (CSF)	G-CSF, M-CSF, GM-CSF and others	Originally identified by ability to make bone-marrow cells differentiate into particular cell type, e.g. neutrophil. Also have effects on mature cells of same lineage, e.g. monocytes, macrophages, neutrophils.
Chemokine	MCP, Eotaxin and many others	Very important in controlling the migration of cells between and within tissues. Also influence function of many cells.
Growth factor	TGF, IGF and many others	Originally identified because of non-immune-related function but may have effects on immune cells.

G-CSF, granulocyte-CSF; M-CSF, macrophage-CSF; GM-CSF, granulocyte/monocyte-CSF; MCP, macrophage chemotactic protein; TGF, transforming growth factor; IGF, insulin-like growth factor.

factors, interferons, tumour necrosis factors, chemokines and growth factors. Growth factors, such as transforming growth factor-β and epidermal growth factor, are now included in the cytokine list although they were not initially identified as having a role in the immune system. Cytokines control many aspects of cell behaviour, including proliferation, differentiation, cell function and leukocyte migration.

The functions of cytokines will be described in detail at the appropriate times when particular mechanisms are being explained. It is important to realise that in the body, cells are never exposed to a single cytokine – they will be exposed to a number of different cytokines, probably produced by a number of different cell types. Different cytokines can either act cooperatively in promoting a response or act antagonistically in inhibiting each other's actions. It is the combination of cytokines to which a cell is exposed that determines the behaviour of the cell.

2.4 The inflammatory response and cell migration

If a pathogen has successfully invaded a tissue, the macrophages in the tissue may recognise the pathogens with one of the receptors described in Section 2.2.2 and attempt to phagocytose and kill the pathogens. Often there are not enough macrophages present in a tissue to phagocytose and remove all the pathogens and therefore the tissue macrophages initiate a response that will bring additional phagocytes, together with a variety of proteins, to the site of infection from the blood. These cells and proteins then help to remove the pathogen. This response is known as the **inflammatory** response. The aim of the inflammatory response is to recruit cells and other factors from the bloodstream into tissues to aid in the remove of pathogens and dead cells or tissue. Leukocytes (white blood cells) are unique in their ability to move throughout the body. They travel through the bloodstream and also have the ability to leave the bloodstream and enter tissue or organs. This ability to move around the body is also referred to as 'cell migration'.

2.5 Cell migration

The movement of cells around the body must be carefully controlled so that the cells go only to where they are required. This control is at two levels: one level is controlling where leukocytes leave the bloodstream; the second level is controlling where the cells go within tissues and organs once they have left the bloodstream. For a single cell, most organs are pretty big places and the cell must go to the right location within the organ or tissue.

Two important factors play an important role in controlling the movement of cells to and within specific tissue sites. **Adhesion molecules** are present on leucocytes and endothelial cells, and interactions between adhesion molecules allow leucocytes to bind to endothelium as part of the process of migrating across the endothelium. **Chemotactic** agents, especially the chemokines, are also important in controlling cell migration. They can act directly on cells and cause them to move in a particular direction or they can act indirectly by altering the expression or binding activity of adhesion molecules.

2.5.1 Adhesion molecules

There are four families of adhesion molecules called **selectins**, **integrins**, **mucin-like vascular addressins** and members of the **immunoglobulin superfamily** (Figure 2.4) and each family contains many members. Different adhesion molecules bind to each other in a specific manner and enable cells to interact with each other. Cell–cell adhesion is controlled both by the expression of particular adhesion molecules and in some cases by the activation status, or actual binding capacity, of the adhesion molecules.

Figure 2.4 Adhesion molecules.

Different adhesion molecules are expressed on different cell types; some are expressed constantly on the cell surface and others are induced by cell activation, e.g. by cytokines.

By altering cell-adhesion molecule expression or activity on endothelial cells or leukocytes, it is possible to control whether particular leukocytes bind to endothelium at a particular tissue site and, hence, the entry of the leukocytes into the tissue.

Selectins are glycoproteins that are lectins, i.e. sugar-binding molecules, some of which are expressed on leukocytes and some on endothelial cells.

Mucin-like vascular addressins are heavily glycosylated proteins and therefore can bind to the selectins. Some are expressed on leukocytes and some on endothelial cells.

Integrins are heterodimeric proteins consisting of an α-chain and a β-chain and are expressed on leukocytes. There are many α- and β-chains and they can pair to give many combinations of integrins with different expression and binding specificity. Some integrins will bind to target molecules only following activation of the leukocyte by various factors.

Immunoglobulin superfamily: these molecules contain immunoglobulin (Ig)-like domains (110 amino acids flanked by an intra-chain disulphide bond) and are the binding target for the integrins. They are expressed on endothelial cells.

2.5.2 Migration of cells from the blood into tissue, and movement of cells in tissues

The process by which cells leave the bloodstream and cross the endothelium to enter into various tissues is called **extravasation**. Although the particular molecules involved may differ in different situations, the fundamental process is the same. Extravasation can be divided into three stages –

rolling, activation and firm attachment, and trans-endothelial migration. Once cells have left the bloodstream they must be guided to the right location within the tissue. The entrance of neutrophils into a site of inflammation is the best understood example and will be described to illustrate the basic steps involved in these processes (Figure 2.5).

1. **Rolling.** Neutrophils, like other leukocytes, normally travel in the centre of the blood flow away from the endothelium. At a site of inflammation vasodilation occurs, slowing down and disturbing the blood flow so that the neutrophils can 'bump' along the endothelium, a process known as rolling. Due to the action of inflammatory mediators, especially TNFα, the endothelial cells are activated to express P-selectin and E-selectin on their surface. These selectins can bind to sialyl-Lewis^x on the surface of the neutrophil, slowing down the neutrophil so that it rolls along the endothelium.

2. **Activation and firm attachment.** The binding of the selectins to the sialyl-Lewis^x is not strong enough for the neutrophil to adhere strongly to the endothelium. Strong attachment requires the binding of the integrin LFA-1 on the neutrophil to ICAM-1 on the endothelium. Before it can bind to ICAM-1, the LFA-1 must change conformation. One of the factors produced in an inflammatory response is interleukin-8 (IL-8), which is a **chemokine**. Chemokines are a group of cytokines with chemotactic and other functions (Table 2.4). Some of the IL-8 produced is held in the extracellular matrix on the endothelial cell surface and can bind to IL-8 receptors on the neutrophil surface. The binding of IL-8 to the neutrophil activates the neutrophil and LFA-1 changes conformation and binds firmly to ICAM-1 on the endothelium.

Table 2.4 Chemokines: families, receptors and functions

Chemokine	Receptors	Cells affected
CXC family		
IL-8	CXCR1, CXCR2	Neutrophils
GRO-α	CXCR2	Neutrophils
IP-10	CXCR3	T cells
CC family		
MIP-1α	CCR-1, CCR-3, CCR-5	T cells, monocytes, DCs
MIP-1β	CCR-1, CCR-3, CCR-5	T cells, monocytes, DCs
MCP-1	CCR2B	T cells, monocytes
RANTES	CCR-1, CCR-3, CCR-5	Eosinophils, monocytes, DCs, T cells
Eotaxin	CCR-3	Eosinophils, monocytes, T cells

The chemokine families are based on the number and pattern of conserved cysteines near the -NH$_2$ terminal of the protein. The CXC family has two cysteines separated by an amino acid. In the CC family the cysteines are adjacent. There are two other families (C) and (CXXXC), which contain two and one members respectively.

Figure 2.5 The stages of neutrophil migration into sites of inflammation. The first stage involves sialyl-Lewis^x on the neutrophil binding to E-selectin and P-selectin on the endothelial cell. Activation of the neutrophil by IL-8 results in a change of conformation of LFA-1 so that it binds firmly to ICAM-1 on the endothelial cell. The neutrophil then squeezes between the endothelial cells and under the influence of chemokines such as IL-8 migrates through the tissue from the tissue to the site of infection.

3. **Transendothelial migration.** Once the neutrophil is firmly attached to the endothelium it squeezes between the endothelial cells, making contact with the basement membrane underneath. This process is poorly understood but involves additional adhesion molecules. Finally enzymes digest the basement membrane, allowing the leucocyte to pass through into the tissue space.

4. **Movement in the site of inflammation.** In the inflamed tissue there will be a gradient of IL-8, with maximum levels at the centre of infection. Neutrophils that have left the bloodstream and entered the tissue will travel along the IL-8 gradient, moving towards increasing concentration of the chemokine so that they will accumulate at the centre of infection.

The way in which other leukocytes cross endothelia, leave the bloodstream and migrate through tissues is essentially the same as for neutrophils, although the adhesion molecules and chemokines may be different for different cell types. Many adhesion molecules and chemokines exist to control adhesion, integrin activation and movement of different types of cells in various tissues. In sites of inflammation other factors such as complement components and prostaglandins can also act as chemoattractants. Tables 2.5 and 6.1 show the main adhesion molecules involved in leukocyte recirculation and migration to sites of inflammation.

Table 2.5 Adhesion molecules involved in migration of leukocytes to sites of inflammation

Adhesion molecule	Cellular distribution	Endothelial ligand
L-selectin	All types of leukocyte	CD34
$\alpha_L\beta_2$ integrin (LFA-1)	T cells, monocytes, macrophages, neutrophils, dendritic cells	ICAM-1, -2, -3
$\alpha_4\beta_1$ integrin (VLA-4)	T cells, monocytes, neutrophils	VCAM-1, fibronectin
CR3	Monocytes, neutrophils, macrophages	ICAM-1
PSGL-1	Neutrophils	E- and P-selectin

2.6 The inflammatory response

As mentioned above, one of the main aims of an inflammatory response is to recruit cells and soluble factors from the bloodstream to help fight off pathogens that have infected a particular tissue site. Four important events occur during an inflammatory response to promote these aims (Figure 2.6):

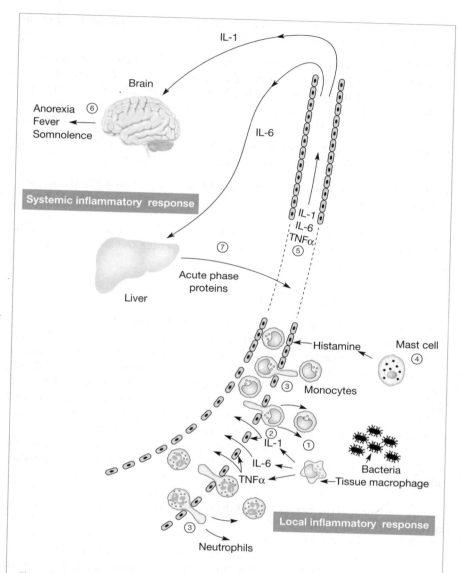

Figure 2.6 Inflammatory responses. Inflammatory responses can be local or systemic. ① Tissue macrophages recognise microbial products. ② The macrophages release cytokines and other inflammatory mediators (IL-1, TNFα, IL-6) that cause vasodilation and increased vascular permeability and have chemotactic effects on monocytes and neutrophils. ③ Monocytes and neutrophils are recruited to the site and there is accumulation of plasma fluid and proteins at the site, causing oedema. ④ Inflammatory mediators can activate mast cells to release further mediators that amplify the response. ⑤ If the local production of cytokines is high enough, the cytokines travel in the blood and affect other organs. ⑥ IL-1 affects the brain, causing fever, anorexia and somnolence. ⑦ IL-6 stimulates hepatocytes to produce acute phase proteins.

- **Vasodilation** causes increased blood flow to the area, increasing the supply of cells and factors.

- **Activation of endothelial** cells lining the blood vessels causes increased expression of adhesion molecules, making the endothelium more 'sticky' to white blood cells so that blood cells can adhere more strongly to the endothelium, thereby promoting the migration of leukocytes from the blood into the tissue.

- **Increased vascular permeability** makes it easier for cells and proteins to pass through the blood vessel walls and enter the tissue.

- **Chemotactic factors** are produced that attract cells into the tissue from the bloodstream.

All of these events are controlled by factors that either are produced by cells involved in the inflammatory response or enter the site of inflammation from the blood.

2.6.1 Activation of macrophages

The first stage in the inflammatory response following infection is recognition of the pathogen and activation of tissue macrophages. The activated macrophages produce a number of factors including prostaglandins, platelet activating factor (PAF) and cytokines. Prostaglandins are a group of small biologically active lipid molecules derived from arachidonic acid. Three of the cytokines produced by the macrophages, interleukin-1 (IL-1), interleukin-8 (IL-8) and tumour necrosis factor-α (TNFα), are important in the inflammatory response. These factors have a number of effects.

TNFα, PAF and the prostaglandins act directly on the endothelium to increase vascular permeability. PAF also causes platelets to release histamine, which is another potent agent for increasing vascular permeability.

IL-1 and TNFα activate endothelial cells lining the blood vessels at the site of infection. This causes the endothelial cells to express on their cell surface molecules that neutrophils in the blood stream can bind to, enabling the neutrophils to leave the bloodstream and enter the tissue. Neutrophils and macrophages ingest and kill bacteria and other microorganisms. The recruitment of neutrophils is also promoted by IL-8, which is chemotactic for neutrophils.

2.6.2 Activation of other pathways during inflammatory responses

A number of other cell types and biochemical pathways can also be activated during an inflammatory response.

Mast cells
Mast cells are distributed throughout the body. They contain many large granules and have similar properties to basophils, which are a type of

mast cells and connective tissue mast cells, which, although sharing most properties, do have some differences (see Chapter 13 for more details on mast cells).

When they are activated, mast cells release the contents of their granules in a process known as mast-cell degranulation. The contents of the granules include histamine, heparin and proteolytic enzymes. These factors result in vasodilation and increased vascular permeability. Activated mast cells also start to synthesise new products, especially prostaglandins and leukotrienes, which are products of the arachidonic acid pathway (Box 2.1). These new products also cause vasodilation and increased vascular permeability and attract neutrophils to the site.

Clotting system

Activation of the clotting system leads to the cleavage of fibrinogen to generate fibrin threads, which form blood clots, and **fibrinopeptides**, which are chemotactic for phagocytes. Formation of a blood clot is important if there has been damage to blood vessels, because the clot can limit the entry of pathogens into the bloodstream and therefore the spread of pathogenic organisms.

Complement system

The complement system is made up of a number of different plasma proteins that play many roles in resistance to infection. It is described in more detail in Chapter 8. In a manner analogous to the clotting cascade, it consists of a series of pro-enzymes and related factors that sequentially activate each other resulting in the production of a variety of biologically active proteins. During inflammatory responses, a complement component called C5a is produced that causes increased vascular permeability. Other complement components, C3a and C5a, can cause mast-cell degranulation, thereby amplifying the inflammatory process.

Kinin system

Kinins are small polypeptides of 9–11 amino acids. They are cleaved from larger plasma proteins called kininogens by specific esterases called kallikreins. The most important kinin in inflammation is **bradykinin**, which causes pain and vasodilation and increases vascular permeability.

All the inflammatory mediators described above cause an increase in blood flow, increase in vascular permeability and chemotactic activity that results in the accumulation of granulocytes and monocytes at the site of inflammation and their activation. The activated macrophages and granulocytes can then begin to remove the pathogenic organisms by the process of phagocytosis.

Prostaglandins (PGs) and leukotrienes (LTs) are small molecules derived from membrane phospholipids. The precursor of PGs and LTs is arachidonic acid, which is produced after cleavage of phospholipids by the enzyme phospholipase. Arachidonic acid can enter the cyclooxygenase pathway, resulting in the formation of PG-G$_2$, which is converted to other PGs and thromboxane B$_2$. Alternatively, arachidonic acid can enter the lipoxygenase pathway, where it is converted to LTA$_2$, which is converted to other LTs. Both prostaglandins and leukotrienes are large families of structurally similar molecules containing 20 carbon atoms. They are chemical messengers and have a number of different functions depending on the particular PG or LT. They are involved in many physiological processes, including control of inflammation, ovulation, parturition, gastric secretion, steroidogenesis and blood pressure.

2.7 The acute phase response

In some cases the inflammatory response will succeed in eliminating the pathogen. In this case the response will be acute (short-lived) and confined to the area of tissue damage. If the pathogen is not eliminated the continued recruitment and stimulation of macrophages will result in a rise in the concentration of the macrophage-derived cytokines in the plasma. These cytokines can affect other organs, particularly the brain and the liver, leading to a systemic response known as an **acute phase response**.

2.7.1 Cytokines and the brain

IL-1 affects the brain, causing fever, somnolence (sleepiness) and anorexia (loss of appetite). Many of the symptoms you feel when you are ill with an infection are due to the actions of cytokines in the brain. Fever is known to have a protective effect in infection, and the replication of some pathogens is inhibited at higher temperatures. Somnolence reduces physical activity and hence energy consumption. Anorexia, by limiting the desire to engage in food-gathering activity, also reduces physical activity. Basically the body is saying, through the actions of these cytokines on the brain, 'rest and concentrate your energies on overcoming this infection'.

2.7.2 Cytokines and the liver

IL-6 has a potent effect on hepatocytes, stimulating them to produce a series of proteins called **acute phase proteins** (APPs). Acute phase proteins are found in the serum at basal (background) levels in healthy normal individuals but rise in concentration following stimulation of the liver. They can be divided into two categories based on the degree to which they increase. The concentration of some acute phase proteins increases only 1.5- to 5-fold while that of others increases 100- to 1000-fold.

APPs that increase 1.5- to 5-fold

- **Fibrinogen.** As mentioned in Section 2.6.2 this is involved in clotting and the generation of fibrinopeptides.

- **Haptoglobulin.** This protein binds to iron-containing haemoglobin and reduces the concentration of iron that many bacteria require for their metabolism, thereby reducing bacterial growth.

- **Complement component C3.** This can be cleaved to generate C3a, which activates mast cells, and C3b, which helps phagocytes recognise pathogens (see Section 2.6.2).

- **Mannose-binding protein (MBP).** MBP is able to bind to mannose-containing sugars on the surface of pathogens and also helps phagocytes recognise pathogens.

APPs that increase 100- to 1000-fold

- **Serum amyloid A (SAA).** This protein inhibits fever and platelet activation. As such, it provides an important negative feedback control loop typical of that seen in many physiological systems.

- **C-reactive protein (CRP).** This protein binds to phosphoryl choline, which is found on the surface of a variety of bacteria, fungi and parasites and is exposed in damaged cells. As described below, it helps phagocytes to recognise bacteria or damaged cells.

The increased serum concentration of the acute phase proteins results in their increased accumulation at the site of inflammation; this is also aided by the increases in blood flow and vascular permeability caused by mediators of the inflammatory response. The acute phase proteins provide additional factors that help in the elimination of infectious agents, especially extracellular dwelling bacteria, yeasts and parasites. Three important APPs are CRP, C3b and MBP, which can act as opsonins to help phagocytes recognise pathogens.

2.8 Opsonins and phagocytosis

The name opsonin derives from Greek, meaning 'prepare for the table'. Victorian biologists likened phagocytosis to eating and hence something that prepared the phagocytes' 'food' for eating was termed an opsonin. The concept of an opsonin is very simple (Figure 2.7): one end of the opsonin recognises and binds to a molecule on the surface of a foreign organism and the other end of the opsonin binds to a receptor on the phagocyte. Engagement of the receptor on the phagocyte then stimulates the phagocytic process.

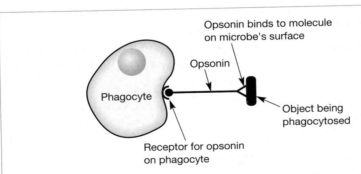

Figure 2.7 Opsonins. Opsonins are proteins that bind to molecules on the surface of microbes and to specific receptors on phagocytes. The binding of the opsonin to the phagocyte receptor activates phagocytosis.

The cells and proteins involved in the inflammatory and acute phase responses are part of the **innate** or **natural** immune system. There are two other components to the innate immune system that will be mentioned here: interferons and natural killer cells

2.9.1 Interferons

Interferons are cytokines that inhibit viral replication in infected cells. They can inhibit viral replication in the cells that produce them, or they can be secreted by the cells and bind to specific receptors on other cells, making them resistant to viral infection (Figure 2.8). The way in which they inhibit viral replication is shown in Box 2.2.

There are three main interferons – IFN-α, IFN-β and IFN-γ. Both IFN-α and IFN-β are produced by many cell types, including macrophages, fibroblasts, lymphocytes, endothelial cells and epithelial cells. IFN-γ is produced by lymphocytes (see Chapter 7) and natural killer cells (see Section 2.9.2). Production of IFN-α and IFN-β is stimulated by microbial products, especially double-stranded RNA, which is made only by viruses and therefore indicates viral infection of a cell. Other stimulators of IFN-α and IFN-β production include products of bacteria, fungi and parasites, and a number of cytokines.

IFNs have many other functions in addition to inhibiting viral replication, two of which are to activate macrophages and natural killer cells (see

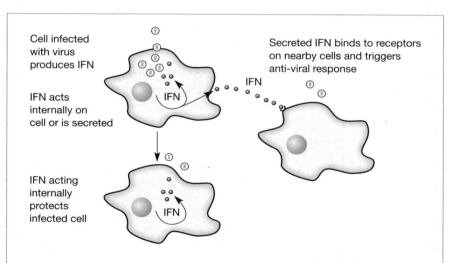

Cell infected with virus produces IFN

IFN acts internally on cell or is secreted

Secreted IFN binds to receptors on nearby cells and triggers anti-viral response

IFN

IFN

IFN acting internally protects infected cell

IFN

Figure 2.8 Interferons. Interferons activate anti-viral responses within the cells that produce them or are secreted and bind to receptors on neighbouring cells, inducing protection against viral infection.

BOX 2.2: ANTI-VIRAL ACTION OF INTERFERONS

The action of interferon α/β involves the induction and activation of two main pathways. An enzyme called 2'-5'-oligoadenylate synthase (2–5(A)-synthase) catalyses the production of 2'-5'-adenylate from ATP. 2'-5'-adenylate activates another enzyme, RNAse, which degrades viral RNA. The second pathway involves the activation of a serine/threonine kinase, called P1 kinase, which phosphorylates the protein synthesis initiating factor elF2, resulting in inactivation of elF2 and inhibition of protein synthesis.

below). They can therefore serve to amplify the innate immune response against the virus and also against other pathogens that stimulate interferon production.

2.9.2 Natural killer cells

Another population of cells that form part of the innate immune system are **natural killer** (NK) cells (Figure 2.9). These cells make up 1–5% of white blood cells. As their name implies, they were first described by their ability to kill other cells, especially tumour cells. It has now been demonstrated that they play an important role in resistance to certain viral, bacterial and protozoan infections, and people without NK cells show increased susceptibility to herpes viruses.

Another important role for NK cells is in the inflammatory response. NK cells enter sites of inflammation, where they can be stimulated by a cytokine

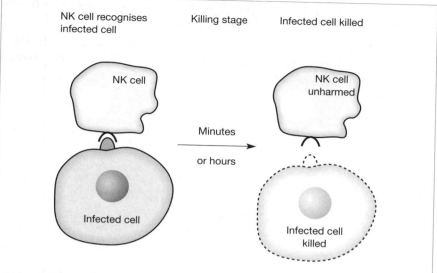

NK cell recognises infected cell | Killing stage | Infected cell killed

NK cell

NK cell unharmed

Minutes

or hours

Infected cell

Infected cell killed

Figure 2.9 NK cell killing. NK cells can recognise cells infected with some pathogens and kill them. They can also kill some tumour cells.

called IL-12, which is produced by activated macrophages. The NK cells are stimulated by IL-12 to produce IFN-γ, which is a powerful activator of macrophages. This provides an amplification loop to maintain macrophage and NK activation. Although NK cells are able to recognise and kill cells infected with certain viruses, it is not clear whether the killing or cytokine-secreting ability of NK cells is most important in resistance to infection.

2.10 Summary

- The first step in responding to infection is to be aware that infection has occurred. Tissue macrophages have receptors that are able to recognise molecules on some pathogens.

- Tissue macrophages can respond to the presence of a pathogen by stimulating an inflammatory response that results in the activation of many cells and protein pathways and the recruitment of phagocytes to the site of inflammation. These phagocytes can remove pathogens and damaged tissue.

- Adhesion molecules and chemokines also control the movement of many cell types into sites of inflammation.

- If the inflammatory response is severe enough, it can also affect the brain, leading to behavioural changes, and the liver, triggering an acute phase response.

- Some of the acute phase proteins can act as opsonins, which help phagocytes phagocytose pathogens.
- Natural killer cells and interferons are other important components of the innate immune system and contribute to the inflammatory response and to protection against viruses.

CHAPTER THREE

Specific immune recognition:

the antibody molecule

Learning objectives

To be introduced to the cells and molecules of the specific immune system. To learn that antibodies are molecules of the specific immune system. To understand how antibodies recognise molecules on pathogens and other structures.

Key topics

● Basic antibody structure
● The nature of binding of antibodies to foreign molecules (antigens)
● Antibody classes: IgG, IgM, IgA, IgE, IgD
● Antibody as a soluble protein and cell-surface receptor.

3.1 Introduction to the specific immune system

Following infection, the first encounter between the pathogen and the infected individual is through the innate immune system, often resulting in an inflammatory response (see Chapter 2). This can have one of two outcomes: either the innate immune system is able to eliminate the pathogen or it is not.

There are many reasons why pathogens are able to avoid elimination by the innate immune system. One reason is that pathogens evolve ways of avoiding being recognised by the cells and opsonins of the innate immune system. A good illustration of this is streptococcal bacteria. Non-virulent streptococcal bacteria are recognised and killed by phagocytes and do not cause disease. Virulent streptococcal bacteria synthesise a waxy polysaccharide (sugar) coat that prevents them being recognised by phagocytes or opsonins of the innate immune system. They cannot be eliminated by the innate immune system and go on to cause disease. Some pathogens, like viruses, are not recognised by the innate immune system because they do not express molecules that are very different from host molecules and therefore cannot be recognised by the innate immune system.

So how do we deal with a pathogen that the innate immune system alone cannot eliminate? The answer is to have help from other cells and molecules with a much better ability to recognise pathogens or their products. Fortunately the immune system has evolved a collection of cells and molecules that are able to distinguish between molecules on pathogens and molecules on host cells, even if the molecules are quite similar. For reasons that will be explained in this chapter, these cells and molecules make up what is known as the **specific** immune system.

The cells of the specific immune system are the lymphocytes, which are a type of white blood cell (Plate 4). There are many different types of lymphocytes with different functions, although morphologically they look the same. One type of lymphocyte, called a B lymphocyte, is responsible for the production of important glycoproteins that are part of the specific immune system. These glycoproteins are called **antibodies** and they have the ability to recognise foreign molecules on the surface of pathogens. Antibodies are also called **immunoglobulins** and the terms antibody (Ab) and immunoglobulin (Ig) mean the same thing. Antibodies are vital for human life and people who cannot make antibodies die of overwhelming infection unless treated with pooled Ig from healthy people. The way in which antibodies recognise molecules derived from pathogens illustrates some of the differences between the innate and specific immune systems and the additional demands that are required of the specific immune system.

3.2 Antibody structure

The basic antibody molecule is depicted as a Y-shaped structure consisting of four protein subunits (Figure 3.1a). The two longer subunits are called **heavy (H) chains** and are identical to each other; they have a molecular mass of 50–75 kDa. The two shorter subunits are also identical to each other; they are called **light (L) chains** and have a molecular mass of about 25 kDa. There are two types of light chain called κ and λ; they are very similar in structure but are coded by different genes. An individual antibody molecule will contain two H chains and two κ-chains *or* two λ-chains; κ- and λ-chains are never seen together in the same antibody molecule. The heavy chains are linked to each other, and to the light chains, by disulphide bridges.

Further analysis of antibody structure shows that both the heavy and the light chains have repeating substructures called **domains** (Figure 3.1b,c). These domains are regions of approximately 110 amino acids within the heavy and light chains and are flanked by intrachain disulphide bridges. The heavy chain of the immunoglobulin molecule has four or five domains and the light chain has two. The domains at the *N*-terminal ends of the heavy and light chains are called the variable domains (V_H and V_L, respectively) because the amino acid sequences of these domains were found to differ from antibody

Figure 3.1 Antibody structure. (a) Each antibody molecule contains two identical larger (heavy) chains and two identical smaller (light) chains. (b) The heavy and light chains contain domains that are structural parts of the proteins, about 110 amino acids in size, flanked by intrachain disulphide bonds. The hinge region contains several prolines giving that part of the molecule flexibility. (c) The N-terminal domains are responsible for binding antigen, and the C-terminal domains of the heavy chain are called the Fc portion of the molecule.

to antibody. The other domains are called constant region domains because they do not differ to the same extent from antibody to antibody. The heavy chain constant domains are called C_H1, C_H2, C_H3 and C_H4; the light chain domain is called C_L. Between the C_H1 and C_H2 domains of the heavy chain is a region containing several prolines; this makes this part of the molecule quite flexible and it is therefore known as the hinge region. X-ray crystallography has revealed that the heavy and light chains fold together to form a molecule that is globular in structure (Plate 5).

3.3 Recognition by antibody – antigens and epitopes

Antibodies bind to molecules that are 'foreign' to the body. These molecules may be on the surface of a pathogen or they may be soluble products such as toxins secreted by pathogens. The molecules that antibodies bind to are called **antigens**. Antigens are nearly always macromolecules; they are usually proteins but they may also be polysaccharides (sugars) or, less commonly, lipids or nucleic acids. An antibody molecule does not bind to the whole of an antigen; it binds to a part of the antigen that is called an **antigenic epitope** (see Figure 3.2). For a protein antigen, an antigenic epitope will be a structural confirmation within the protein ranging from 8 to 22 amino acids in size. Epitopes recognised by antibody can be linear or

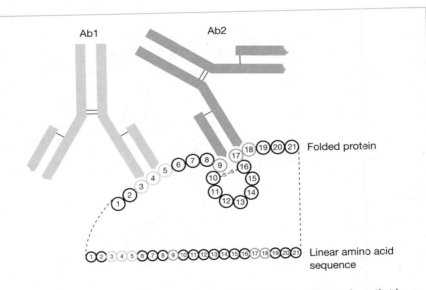

Figure 3.2 Linear and conformational epitopes. Ab1 is binding to a linear epitope that is composed of amino acids in a continuous sequence in the protein amino acid sequence (amino acids ③, ④ and ⑤). Ab2 is binding a conformational epitope formed by the folding of the protein antigen. The amino acids forming the structural epitope (⑨, ⑰ and ⑱) do not occur in a linear sequence in the amino acid sequence of the protein antigen. Note that in reality antigenic epitopes are formed by 5–15 amino acids and not the three depicted here.

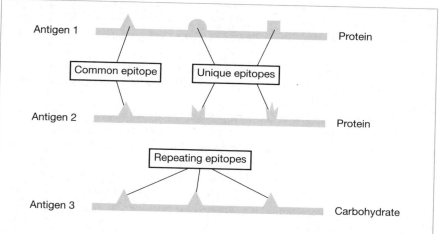

Figure 3.3 Repeating and unique epitopes. Antigen 1 has three different epitopes depicted by the triangle, circle and square shaped protrusions. Antigen 2 has one epitope (triangle) that is shared by antigen 1 and two unique epitopes. Antigen 3 has repeats of the same epitope; this is typical of polysaccharide antigens.

conformational (Figure 3.2). A linear epitope is a conformation on the antigen formed by a continuous sequence of amino acids. By contrast, a conformational epitope is formed by the folding of the protein (Figure 3.2). Given that some proteins have an M_r of over one million and contain thousands of amino acids, a single antigen can potentially have a number – possibly several hundred – of different epitopes. Although it might be expected that a large protein could theoretically form a large number of epitopes, in reality this is not the case. An experimental antigen, hen-egg lysozyme, is a protein of M_r 14 000 and appears to have eight different antigenic epitopes. Most antigens are thought to have a similar number of epitopes in relation to their M_r. For most antigens each epitope on one molecule of antigen will be different, although some polysaccharides have many repeats of the same epitope (Figure 3.3).

3.3.1 The antigen-binding site of antibodies

The parts of the antibody molecule that bind to antigenic epitopes are called the antigen-binding sites and are located in the variable domains of the H- and L- chains (Figures 3.2 and 3.4). Because they are responsible for antigen binding they have also been called the antigen-binding fragments, or FAbs, of the molecule (Box 3.1). The antigen-binding sites are formed by the folding of the variable domains of the heavy and light chains (Figure 3.4). Antibodies against different antigens have different amino acid sequences in their variable regions. More detailed analysis shows that the variability is confined to three regions, each of 5–15 amino acids in length, within each heavy and light chain (Figure 3.4); these regions are therefore called the **hyper-variable** regions are the antibody. The parts of the variable

BOX 3.1: FRAGMENTS OF ANTIBODIES

Much of the terminology about the different parts of antibodies comes from early (1950s and 1960s) experiments to try to elucidate the structure of antibodies. It was known that antibody (the IgG fraction of serum) had a molecular weight of 150 000 but it was not known whether Ig consisted of one protein chain of M_r 150 000 or was made up of smaller units.

The approach used to address the problem was to see whether the antibodies could be broken down into smaller subunits. One approach used enzymes to break down antibodies. Brief incubation of Ig with the enzyme **papain** (found in the papaya latex) was found to produce two fragments, one of M_r 45 000, which was called **FAb** (for 'fragment antigen binding') because this fragment could still bind antigen, and one of M_r 50 000 called **Fc**, which crystallised upon storage at 4 °C (Fc stood for 'fragment crystallisable'). Another enzyme, pepsin, generated a single fragment of M_r 100 000, which became called **F(Ab)$_2$**. Another reagent used was 2-mercapto-ethanol (2-ME), which breaks interchain disulphide bonds. Treatment of Ig with 2-ME produced two fragments of M_r 50 000 and M_r 25 000, which were the heavy and light chains respectively.

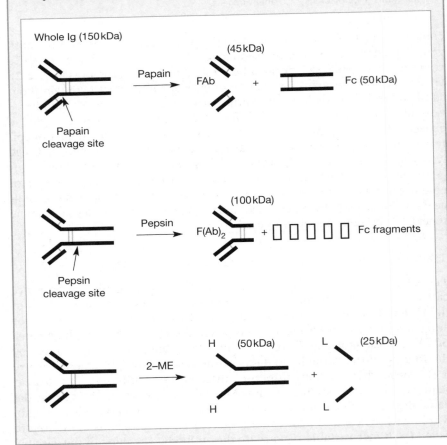

Eventually the structure of IgG was deduced and the basis for the different patterns of cleavage explained as shown in the figure. This work led to the award of the Nobel Prize to Rodney Porter in the UK and Gerald Edelman in the USA.

regions between the hyper-variable regions do not differ so much between antibodies and have been called the **framework** regions. They were so named because they contribute to the overall structure of the antibody and provide a framework on which the hyper-variable regions sit. Although the hyper-variable regions are equally spaced in the linear amino acid sequence of the variable region, when protein folding is taken into account it can be seen that the three hyper-variable regions are brought together at the end of the molecule (Figure 3.4). Furthermore, the heavy and light chains fold together so that their hyper-variable regions form a single surface. This surface forms the antigen-binding site and interacts with the antigenic epitope (Figure 3.4 and Plate 6). Different antibodies will have antigen-binding regions with different shapes, as depicted in Figure 3.5. These shapes determine whether an antigenic epitope will bind to the antibody.

Hyper-variable regions and complementarity-determining regions
The terms 'hyper-variable region' and 'complementarity-determining region' mean the same thing but it is not always easy to understand why. For an antibody to bind to an antigenic epitope, the shape of the epitope must fit

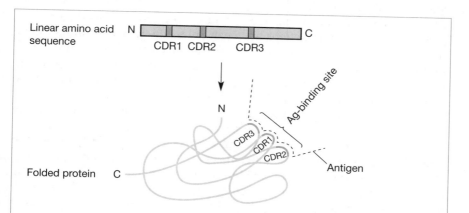

Figure 3.4 Hyper-variable regions of antibody chains. These fold to form the antigen-binding site. The three hyper-variable regions are spaced apart in the linear amino acid sequence of the variable region of the heavy and light chain proteins. However, when protein folding occurs the three hyper-variable regions come together to form a single antigen-binding site. This occurs for both H- and L-chains so that the H + L combination forms a single antigen-binding site composed of six hyper-variable regions – three from the H-chain and three from the L-chain.

Figure 3.5 Different antibodies bind different antigenic epitopes. An antigen is shown schematically with an antigenic epitope depicted as a triangle. Ab1, Ab2 and Ab3 have different-shaped antigen-binding regions depicted by a ∧, ∩ or ⊓ shape on the ends of the arms of the antibody. Only Ab1 will bind the antigenic epitope.

that of the antibody. Another way to say this is that the antigen-binding site of the antibody is **complementary** in shape to the antigenic epitope (Figure 3.4). It is the hyper-variable regions of an antibody that determine whether it is complementary to an antigenic epitope, and so the hyper-variable regions of an antibody are also called the **complementarity-determining regions**, or CDRs for short.

3.3.2 The nature of antibody–antigen binding – specific recognition

The chemical interactions between antibody and antigen are non-covalent. Four types of non-covalent interaction are involved: hydrogen bonds, electrostatic forces, van der Waals forces and hydrophobic forces (Figure 3.6). These non-covalent interactions are weak unless the two molecules forming the bond are very close together in molecular terms; thus a lot of these interactions are required for strong binding of antibody to antigen. However, as the molecules come close together, electrons of atoms in the antigen and antibody repel each other because they are negatively charged. The strength of binding between an antibody binding site and an antigenic epitope is therefore determined by the net balance between the attractive and repulsive forces. This is why there must be a good fit between the antibody-binding site and the antigenic epitope for strong binding to occur (Plate 6 and Figure 3.7). Weaker binding may occur if the fit between the antibody-binding site and the antigenic epitope is not so good. The strength with which the antigen-binding site of an antibody binds to an antigenic epitope is known as the **affinity** of the antibody for the antigen.

In many situations one amino acid in the antigenic epitope can determine whether it will bind antibody. The antigen to which the antibody is binding in Plate 6 is called hen-egg lysozyme; this is a common antigen used for experimental purposes in studying immune responses and antigen–antibody interaction. One amino acid, a glutamine, in the antigenic epitope of hen-egg lysozyme forms strong hydrogen bonds with the binding site of the antibody. Turkey-egg lysozyme is almost identical to hen-egg

Figure 3.6 Forces involved in antigen binding by antibody.

Antigen A

Good fit gives high-affinity (strong) binding

Antigen B

Poor fit gives low-affinity (weak) binding

Antigen C

No fit gives no binding

Figure 3.7 Antibodies can bind antigenic epitopes with different affinities. The same antibody is shown interacting with three antigenic epitopes. With antigen A there is a good fit between the antibody-binding site and the antigenic epitope, giving high-affinity (strong) binding. There is a poor fit between the antibody and the epitope on antigen B, resulting in some binding of low affinity (weak binding). The antibody-binding site does not fit the epitope on antigen C, so there is no binding.

lysozyme except that it has substituted another amino acid for the glutamine and as a result the antibody will not bind to turkey-egg lysozyme. Therefore antibodies are capable of binding to one protein but not to another protein that differs in amino acid sequence by only one amino acid. This means that even if a protein on the surface of a pathogen differed from an individual's own protein by one amino acid, the antibody would still be able to bind to the pathogen but not the host cells. This provides an incredibly powerful way of distinguishing between foreign and self molecules.

3.3.3 Affinity and avidity of antibody–antigen interactions

These two terms often cause confusion. Antibodies have at least two anti-gen-binding sites, each made up of a heavy-chain/light-chain pair of polypeptides. As mentioned above, the strength with which an individual binding site of an antibody binds its epitope is called the **affinity** of binding. However, if an antibody is using both binding sites to bind to two epitopes on the same particle the overall strength of the binding is increased and the total strength of this binding is called the **avidity** of the antibody for the antigen (Figure 3.8). Avidity is more than a simple sum of the affinities of each antibody-binding site for its antigenic epitope. To break the interaction between multiple binding sites it is necessary to break the binding at every binding site at the same time and this requires much more energy than breaking the binding at a single binding site.

A way that is often used to illustrate the difference between affinity and avidity is to imagine that you are hanging on to a high wire by one or both hands. If you were hanging by one hand, the strength with which you could hang on to the wire would be determined solely by the strength of your grip. The strength of grip can be likened to the affinity of a single antibody-binding site for its antigenic epitope. If sufficient force was applied to loosen your grip on the wire you would immediately fall to whatever fate awaited you. Now imagine that you are hanging on to the wire by both hands. The same force that could loosen the grip of only one hand would not be strong enough to loosen the grip of both hands. Therefore the overall strength by which you are hanging on to the wire is much greater; this rep-resents the avidity of antibody-binding to an antigen using two or more binding sites.

(a) Antigen with
    non-repeating
    epitope

(b) Multiple repeats of antigen
    on particle, e.g. microbial
    cell wall

(c) Antigen with
    repeating
    epitope

Figure 3.8 Affinity and avidity. (a) With most soluble protein antigens, such as secreted bacterial toxins, each epitope occurs only once on each protein molecule, although there may be a number of different epitopes on each protein molecule – these are called non-repeating epitopes and the strength of binding of the antibody is determined solely by the affinity of the antibody binding site for the antigenic epitope. (b) If several copies of the protein are present on the surface of a particle, such as a bacterium, each antibody-binding site can bind to an epitope, and therefore the overall strength of binding of the antibody to the **particle**, or avidity of binding, is determined by the number of binding sites and the affinity of each binding site. (c) Some antigen molecules, especially carbohydrates, have more than one copy of an epitope and therefore both antibody-binding sites can bind to the antigen, resulting in increased avidity as in (b).

We have talked about antibodies as opsonins. To fulfil this role the antibody must bind to the particle that is to be phagocytosed and to the phagocyte. Antibodies bind to the particle by specifically binding to antigens on the surface of the particle, which could be a bacterium or virus. Phagocytes have receptors, called Fc receptors (FcRs), on their cell surface that can bind to the Fc portion of antibody molecules that have bound antigen. This enables the phagocyte to phagocytose the whole complex of the antibody and particle (Figure 3.9).

Although opsonisation is a very important role of antibodies, it is not the only function of antibodies. (See Chapter 8 for more details on antibody function.) The many different functions require different properties within the Fc portion of the antibody molecule and therefore different antibodies exist with different Fc regions. These different types of antibody are called **classes** of antibody. There are five different classes of antibody in humans, called IgM, IgG, IgA, IgE and IgD. Slightly different variations of IgG and IgA exist, which are called **subclasses** of antibody. There are four subclasses of IgG and two of IgA. Each class or subclass of antibody has a unique set of functions determined by the Fc part of the molecules (except for the two subclasses of IgA, which appear to have identical function). However, the specificity of antibody for antigen is related not to its class but to the shape of the antigen-binding site. Different classes of antibody can have the same antigen specificity.

3.4.1 Immunoglobulin G (IgG)

This is the most abundant antibody in serum (see Table 3.1) and exists as the basic 2H + 2L chain antibody molecule (Figure 3.10). The four subclasses of IgG in the human are IgG1, IgG2, IgG3 and IgG4, all having the same 2 heavy + 2 light chain structure (Table 3.1). The different IgG subclasses have slightly different structures and functions, although they are

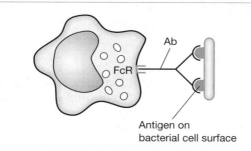

Figure 3.9 Antibody as an opsonin. The antigen-binding sites of the antibody bind to antigens on the bacterium and the Fc part of the antibody binds to the Fc receptor (FcR) on a phagocyte, stimulating phagocytosis of the bacterium.

Table 3.1 Antibody classes

Class	Heavy chain	Molecular mass (kDa)	Serum concentration (mg/ml)	Serum half-life (days)
IgM	μ	900	1.5	10
IgG1	γ1	150	9	21
IgG2	γ2	150	3	20
IgG3	γ3	165	1	7
IgG4	γ4	150	0.5	21
IgA1	α1	160	3	6
IgA2	α2	160	0.5	6
IgE	ε	190	5×10^{-5}	2
IgD	δ	185	0.03	3

closely related to each other. IgG1 is the most common of the IgG sub-classes, comprising about 70% of total IgG.

3.4.2 Immunoglobulin M (IgM)

IgM is the earliest antibody to be produced after first contact with a new antigen. Structurally it consists of a pentamer of five Ig molecules joined together by disulphide bonds and an extra protein called the joining or J-chain (Figure 3.10). The H-chains of IgM differ from those of IgG in having four constant domains instead of the three in IgG. The J-chain of IgM has a molecular weight of 15 000 and is coded by a gene on a separate chromosome from the genes coding immunoglobulin heavy and light chains. It binds through disulphide linkages to the heavy chain of two of the five Ig molecules that make up IgM (Figure 3.10). The J-chain is required for the proper polymerisation of IgM into pentamers and in the absence of J-chain pentamers are not formed. IgM potentially has ten antigen-binding sites although steric hindrance means that in practice it can usually use only up to six of them at once.

3.4.3 Immunoglobulin A (IgA)

IgA is present in the serum and is also the main class of antibody found in various secretions such as mucus in the intestinal and respiratory tracts, saliva, sweat, breast milk and colostrum. The two subclasses of IgA in the human, IgA1 and IgA2, appear to have the same functions. IgA has different structures depending on whether it is in serum or secretions. In serum, IgA adopts the basic 2 heavy + 2 light chain Ig structure (Figure 3.10). IgA in secretions has a different structure and consists of two Ig molecules joined together by a J-chain (as in IgM) and an additional protein called the secretory piece. The secretory piece is not actually made by the antibody-

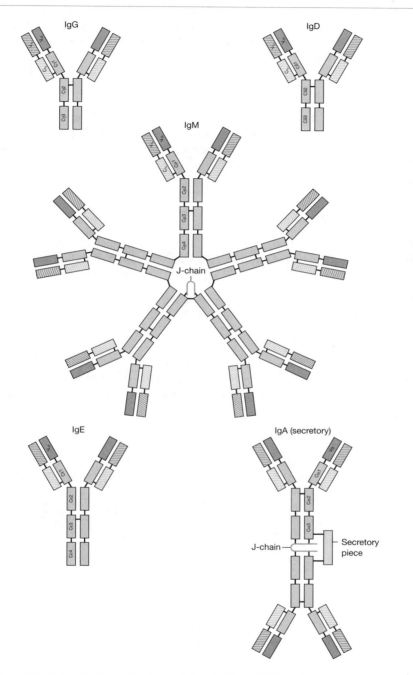

Figure 3.10 Antibody classes. The heavy chains of IgM and IgE have four C_H-domains compared with the three of other Ig classes. Secretory IgA is shown. IgA also exists as a monomer in serum, without the J-chain and secretory piece.

producing cell but is added to the IgA in a special way (see Section 7.4). The secretory piece helps the transport of the IgA into secretions and also helps to protect the IgA from breakdown by proteolytic enzymes that are found in secretions.

3.4.4 Immunoglobulin E (IgE)

IgE is present at the lowest concentration of all antibody classes in serum. It exists as a monomer consisting of the basic two heavy and two light chain Ig structure. Like IgM, its heavy chain also has four C-domains (Figure 3.10). IgE has very special functions and is involved in asthma and allergy (see Chapter 13).

3.4.5 Immunoglobulin D (IgD)

IgD generally has a low serum concentration and is unstable in serum, being quickly degraded by serum plasmin. It exists as a monomer of the basic Ig structure (Figure 3.10) but the functional significance of serum IgD is not clear.

3.5 Antibody can be secreted or expressed on the cell surface of B lymphocytes

The cells that make antibody are called plasma cells (Plate 4). These are derived from a type of white blood cell called a B lymphocyte. B lymphocytes themselves are produced in the bone marrow and then circulate

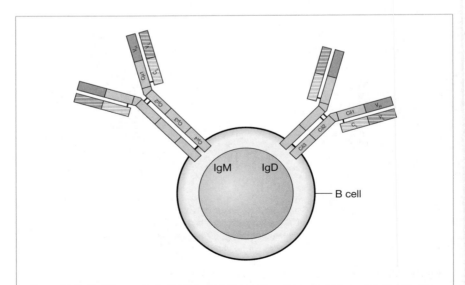

Figure 3.11 B cells express both IgM and IgD as cell-surface membrane proteins. The V_H region is the same on both IgM and IgD and both the IgM and IgD molecules contain identical L-chains.

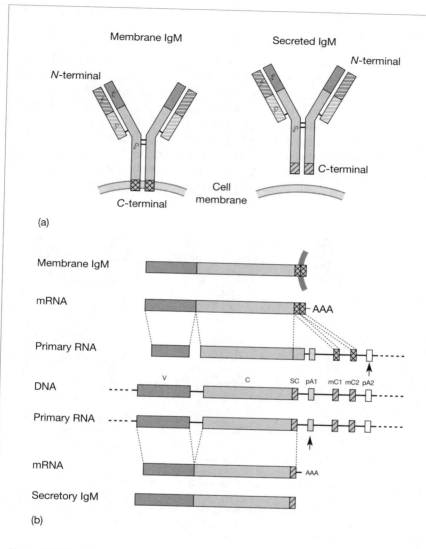

Figure 3.12 Production of membrane and secreted IgM. (a) A B cell can both express IgM on its cell surface and secrete IgM. The surface and secreted IgM are identical except for the *C*-terminal ends of their H-chains (shown by different shading). The *C*-terminal end of membrane IgM contains many hydrophobic amino acids that anchor it in the cell membrane; secreted IgM does not have these. Both secreted and membrane Ig H-chain associate with the same light chain made by the B cell. (b) A simplified diagram of the genes coding for IgM show alternative polyadenylation sites, pA1 and pA2, and genes coding for the secretory component (SC) or membrane component (MC1 and MC2) of the IgM H-chain. The primary RNA transcript is the same for both secretory and membrane IgM. Polyadenylation at pA1 and RNA splicing leads to mRNA that codes for the secreted IgM, while polyadenylation at pA2 and different RNA splicing lead to mRNA coding for membrane IgM.

through the bloodstream and spleen, lymph nodes and other lymphoid tissue (see Chapter 6).

B lymphocytes express the immunoglobulin molecules IgM and IgD on their cell surface as integral membrane proteins (Figure 3.11). Putting Ig into the cell membrane rather than secreting it is achieved by differential processing of the heavy chain RNA (Figure 3.12). All the IgM and IgD molecules on the surface of any one B cell have the same heavy chain variable regions and are associated with the same light chain; only the heavy chain constant regions differ between the IgM and IgD. Each B cell therefore has one set of heavy and light chain variable regions and one antigen specificity. Different B cells will have different sets of heavy and light chain variable regions and therefore each B cell has a different antigen specificity (Figure 3.13).

Figure 3.13 Different B cells express membrane Ig molecules with different antigen specificity. All the Ig molecules expressed by a single B cell have the same H- and L-chains. However, different B cells will have different V_H and V_L regions (indicated by different shading and shapes at the end of the V regions) and therefore different B cells will have different antigen specificity.

The membrane Ig on the B lymphocyte surface forms a signalling complex with two other proteins called Igα and Igβ. This signalling complex actually consists of eight chains: two identical Ig heavy chains (μ or δ), two identical Ig light chains (κ or λ) forming the Ig molecule, and two dimers each consisting of an Igα and a Igβ chain (Figure 3.14).

Therefore antibody can exist in two forms: a soluble form that has a number of biological activities depending on the class of antibody (see Chapter 8) and an integral cell-membrane protein on the surface of B lymphocytes that enables B lymphocytes to recognise a specific antigen and respond to it.

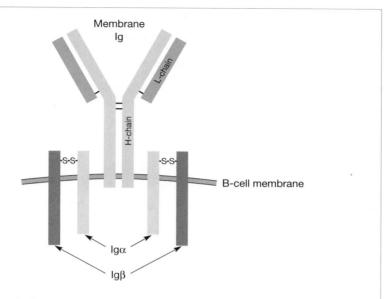

Figure 3.14 The B-cell receptor complex. Each membrane Ig molecule on the surface of a B cell is associated with two heterodimers; each heterodimer consists of an Igα chain linked by disulphide bonds to an Igβ chain. The Igα and Igβ chains are responsible for initiating intracellular signalling when the Ig binds antigen.

3.6 Summary

- Antibodies are a group of molecules with recognition and biological functions. Each antibody has two identical heavy chains and two identical light chains.

- The *n*-terminal parts of the heavy and light chains are called variable regions. These fold together to form the antigen-binding site, which binds to antigens on pathogens or other foreign objects. Antigens are most often proteins but can be carbohydrates or, less commonly, lipids. Antibodies bind to small parts of antigens called antigenic epitopes.

- Recognition of antigens by antibodies differs from recognition by the innate immune system in that antibody-binding is specific. This means that one antibody molecule will bind to some, but not all, antigenic epitopes. Different antibodies will bind to different epitopes, i.e. they will have different specificities.

- Because each antibody will bind to only a very small proportion of the total number of antigenic epitopes, lots of different antibody specificities are required to recognise all the different antigens.

- The *c*-terminal part of the antibody molecule is called the constant region. Different classes of antibody exist with different constant regions. The different classes of antibodies are called IgM, IgG, IgA, IgE and IgD and they have different functions.

- Immunoglobulin can be secreted as a soluble protein or can be put into the cell membrane of B lymphocytes, where it can be part of a cell signalling complex.

CHAPTER FOUR

T lymphocytes and MHC-associated recognition of antigen

Learning objectives

To know the structure of the MHC and the roles of MHC molecules in antigen presentation to T cells. To know how T cells recognise antigen. To learn about antigen processing. To appreciate the unique polymorphism of class I and class II MHC molecules and the advantage of this.

Key topics

- T lymphocyte subsets
- Major histocompatibility complex (MHC)
  - ○ Genetic organisation of the MHC
  - ○ MHC gene products – class I and class II MHC proteins
  - ○ Polymorphism of MHC genes
  - ○ Expression of MHC proteins
- Recognition of antigen by T cells – structure of the T cell receptor for antigen
- Antigen processing and presentation by class I and class II MHC

4.1 Overview of T lymphocyte subsets

The previous chapter described one of the important recognition molecules of the specific immune system – the antibody molecule – which can be secreted in soluble form or be present on the surface of B lymphocytes. Through antibody on their cell surface, B lymphocytes can recognise antigen and this triggers a cellular response in the B lymphocyte. In this chapter we will describe a different type of lymphocyte that can also recognise antigen but in a different way from B lymphocytes. These cells are the **T lymphocytes.**

The T in T lymphocyte stands for **thymus**-derived. The thymus is a bi-lobed organ situated in the mediastinum above the heart (see Chapter 11). How the importance of the thymus in T cell development was discovered is described in Box 4.1. T lymphocyte precursors, like those of B lymphocytes, originate in the bone marrow. Whereas B cells complete their maturation in the bone marrow, T lymphocyte precursors migrate to the thymus where they develop into mature T lymphocytes (T cell development is described in Chapter 11). The mature T lymphocytes then leave the thymus and circulate through the bloodstream and lymphoid tissue (see Chapter 6).

In addition to helper T cells there is another major type of T cell, which is also produced in the thymus and has different functions from helper T cells. Both of these cells are called T cells or T lymphocytes. The two types of T cell are distinguishable phenotypically by the expression of mutually exclusive molecules on their cell surface. Helper T cells express a molecule called CD4 on their cell surface and are therefore called **CD4 T cells**. Cells

BOX 4.1: DISCOVERY OF THE IMPORTANCE OF THE THYMUS

Prior to the work of Jacques Miller in the late 1950s and early 1960s, the role of the thymus was obscure because its removal from adult mice or humans had no apparent adverse effects. Miller was working for the Chester Beatty Cancer Research Institute investigating the role of the thymus in the development of leukaemia caused by the injection of a virus, called Gross leukaemia virus, into mice. (The virus was called Gross not because it was particularly disgusting but because it had recently been isolated by Ludwik Gross in the US.) The virus caused leukaemia when injected into neonatal (newborn) mice. However, if the mice were thymectomised (i.e. the thymus was removed) when they were weaned (at four weeks of age), the mice did not develop leukaemia. Therefore the thymus was essential for the development of leukaemia in this model.

Miller wished to test the effect of thymectomy *before* the neonatal mice were inoculated with virus. He therefore neonatally thymectomised many mice of different strains for these experiments. What he observed was that the neonatally thymectomised mice grew well until they were weaned when, whether they had been infected with virus or not, they suffered a wasting disease and died early.

Histological examination showed that the mice had reduced lymphocytes in the blood and lymphoid tissue, and liver lesions suggestive of viral infection. Further immunological studies revealed that the neonatally thymectomised mice could not reject skin grafts from other strains of mice, or even other species, showing that the mice were profoundly immunodeficient. Further evidence showed that if the mice were kept in germ-free conditions they did not suffer the wasting disease and survived for much longer, indicating that early death was due to infection. These studies, on the completely different subject of the development of leukaemia, were among the first to establish that the thymus did have an important role in immune function.

of the other T lymphocyte subset express a different molecule called CD8 on their cell surface and are called **CD8 T cells**. T lymphocytes in the periphery express either CD4 or CD8 but not both.

4.2 The T cell receptor for antigen

Although there are similarities, T cells recognise antigen in a fundamentally different way from B cells. Both CD4 and CD8 T cells use a receptor called the T cell receptor (TcR) for recognising antigen. The TcR is related to, but is different from, antibody, and the genes coding for the TcR are on different chromosomes from those coding for antibody. The TcR consists of two glycoprotein chains called α and β (Figure 4.1). The α-chain has a molecular size of 40–50 kDa and the β-chain 35–47 kDa. Both chains show a typical Ig-like domain structure showing that TcR and Ig evolved from a common gene. Similar to Ig, each chain of the TcR has a constant domain near the cell membrane and variable domain farthest from the cell membrane (Figure 4.1). The variable domains contain hyper-variable regions and, again like antibody, the TcR adopts a globular structure and the hyper-variable domains of the α- and β-chains fold to provide an antigen-binding site. The TcR is never secreted like antibody but exists only as a receptor on the surface of T cells.

Just as antibody on the B lymphocyte cell surface is associated with intracellular signalling molecules, the TcR is associated with a group of proteins with signalling function called CD3. There are five CD3 proteins called gamma (γ), delta (δ), epsilon (ε), zeta (ζ) and eta (η). The ε chain can form

Figure 4.1 Structure of T cell receptor for antigen (TcR). The TcR consists of two glycoprotein chains, an α-chain of M_r 40–50 kDa and a β-chain of M_r 35–47 kDa. The Cα, Cβ, Vα and Vβ domains are structurally similar to the Ig domains. The α and β chains are associated with the CD3 complex of γ, δ, ε, and ξ chains (left).

heterodimers with γ or δ and the ζ chain can form a homodimer with another ζ chain or a heterodimer with a η chain (see Figure 4.1). All CD3 complexes with the TcR contain $\gamma\varepsilon$ and $\delta\varepsilon$ dimers, about 90% contain $\zeta\zeta$ dimers and 10% contain $\zeta\eta$ dimers. The CD3 complex is not involved in antigen recognition but initiates intracellular signalling pathways when TcRs recognise antigen.

T cells follow some of the same rules as B cells: different T cells will recognise different antigens because they have TcRs with different variable regions on their cell surface, but each T cell will express only one TcR specificity. Both CD4 and CD8 T cells use the same genes to produce their TcR. However, T cell recognition of antigen differs in one fundamental aspect from the way in which antibody recognises antigen. T cells do not recognise free antigen in the way antibody can. They recognise antigen that is associated with molecules on the surface of cells called **major histocompatibility complex (MHC) molecules**. Before describing how T cells recognise antigen, it is necessary to know about the major histocompatibility complex.

4.3 The major histocompatibility complex

The terminology surrounding the major histocompatibility complex tends to make the whole subject sound more complicated than it actually is. Therefore some of the terminology will be explained before describing the MHC in more detail.

4.3.1 Terminology of the MHC

Major histocompatibility complex (MHC)
The term major histocompatibility complex actually refers to a region of DNA spanning some 4 Mbp (base pair) and containing over 100 genes. It is located on chromosome 6 in the human and chromosome 17 in the mouse. The organisation of the genes in humans is shown in Figure 4.2. All vertebrate species have an MHC, and therefore the term MHC can be used in reference to any species. Each species also has a unique name for its MHC; in humans the MHC is also called HLA (**h**uman **l**eucocyte **a**ntigen) and in mouse it is called H-2. Therefore, in the human, the terms MHC and HLA mean the same thing. The origin of the terminology surrounding the MHC is described in Box 4.2.

MHC gene products
The proteins coded for by the genes of the MHC are called MHC proteins, MHC molecules or MHC products; these terms mean the same thing and are used interchangeably. Because they were initially identified as antigens involved in transplant rejection, MHC proteins are also often referred to as MHC antigens.

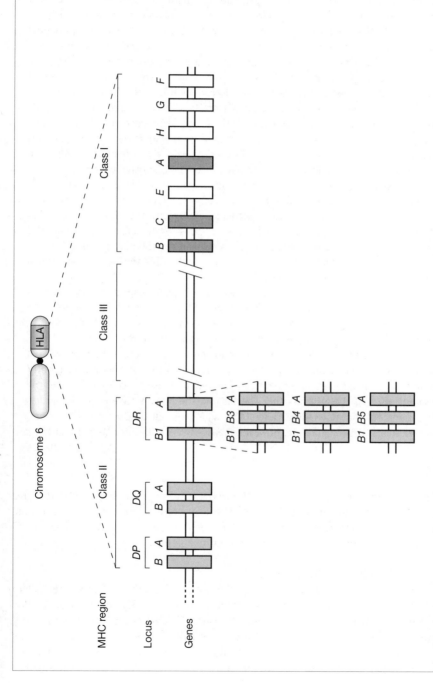

Figure 4.2 Organisation of HLA genes. The class III region contains numerous diverse genes. Each of the *HLA-DRB* genes is polymorphic. The organisation of the *DRB* genes is such that, depending on the form of gene inherited, an individual possesses an *HLA-DRB1* allele alone or a *HLA-DRB1* allele plus one of *DRB3*, *DRB4* or *DRB5*. Each of the *DRB1*, *DRB3*, *DRB4* and *DRB5* loci are polymorphic.

BOX 4.2: TERMINOLOGY OF MHC IN DIFFERENT SPECIES

The major histocompatibility complex was first discovered in mice, where it was called H-2. The H stood for histocompatibility. The word histocompatibility is derived from the Greek words '*Histos*' meaning tissue and '*Compatibilitos*' meaning compatible. Early in the twentieth century tumour biologists were transplanting tumours between mice. They observed that the tumour grafts suffered one of three fates: they grew and killed the mouse, they were rejected slowly (>50 days) or they were rejected rapidly (<20 days). When inbred strains of mice were used it was found that the three fates of the tumour grafts were observed between the different strains. Two strains of mice that could exchange tumour grafts without destruction of the graft were said to be histocompatible, i.e. their tissues could co-exist with each other. By contrast two strains of mice that rejected each other's tumours were said to be histo-incompatible, i.e. their tissues could not co-exist with each other. If the rejection was rapid the strains were said to show major incompatibility and if the rejection was slow the strains were said to show minor incompatibility. Using classical genetic breeding studies it became clear that rapid rejection segregated as a single locus and this was originally called the major histocompatibility locus. Later it was realised that the locus contained many genes and the word 'locus' was replaced with 'complex' – hence the term major histocompatibility complex and the abbreviation MHC. Because it was the second histocompatibility locus to be identified in the mouse it was called H-2 (standing for histocompatibility-2).

It was natural to look for a similar system in humans, especially with the possibility of human transplantation. Indeed an MHC was discovered in humans and called HLA, standing for **h**uman **l**eucocyte **a**ntigen because the MHC **proteins** in humans were first identified as transplantation antigens on white blood cells.

Therefore, although there was a logic to the nomenclature in both mouse and human, not everyone spoke the same language. When the rat MHC was called RT1 things threatened to get out of hand. Fortunately there is now consensus about the nomenclature of the MHC in other species and the human-style terminology has been adopted. Thus dog MHC is called DLA, pig MHC is called SLA (swine LA) , chimpanzee MHC is called ChLA, etc. However, mouse MHC is still called H-2 and rat RT1.

MHC locus

Initially it was thought that the MHC was one gene, or locus, but it quite soon became apparent that there were many genes. The term 'MHC locus' is now used to describe the region of DNA coding for an individual MHC product; each locus may contain one or more genes.

Classes of MHC gene or product

The MHC contains over 100 genes. However it is clear that some of the genes are closely related to each other and their products perform a similar function. On this basis the genes of the MHC, and the proteins coded by them, have been divided into three classes called class I, class II and class III. The class I and class II MHC molecules are related to each other but fall into two distinct families of proteins. They are nearly all cell surface

Table 4.1 MHC class III gene products

Gene product	Function	
Steroid 21 hydroxylase	Enzyme	
C2	Complement component	
C4	Complement component	see Chapter 8
B	Complement component	
Hsp 70 (heat shock protein)	Intracellular trafficking of proteins	
TNFα	Inflammatory cytokine	
TNFβ	Inflammatory cytokine	

proteins and are involved in T cell recognition of antigen (see below). Class III MHC molecules are a diverse collection of proteins that have many immune-related functions but are not related to class I or class II MHC or necessarily to each other. Table 4.1 shows some of the major types of MHC class III products.

4.3.2 Class I MHC

Structure of class I MHC molecules

Class I MHC molecules are glycoproteins that are expressed on the cell surface of most nucleated cells (Figure 4.3). They consist of two protein chains. The longer chain, called the α-chain, is coded for by the MHC class I genes and has a molecular size of 45 kDa. The α-chain is non-covalently linked to a 12 kDa chain called β_2-microglobulin (also referred to as β_2-m), which is coded for by a gene on a different chromosome (Chr 15 in humans) from that for the MHC.

The α-chain has three external domains, called α_1, α_2 and α_3, that are homologous to immunoglobulin domains. The α_1 domain is located at the N-terminal end of the molecule while the α_3 domain is situated closest to the cell membrane. Like most cell surface molecules the α-chain has a short hydrophobic region spanning the membrane and a short cytosolic region. There are no known signalling functions associated with class I MHC. β_2-microglobulin has one Ig-like domain and is located exclusively on the outside of the membrane; it has no transmembrane or intracellular regions. Diagrammatically class I MHC molecules can be depicted as two sets of domains (the α_1 and α_2) sitting on top of the other two (α_3 domain and β_2-microglobulin) (Figure 4.3).

The detailed molecular structure of class I MHC was determined by X-ray crystallography in 1987 and shows that the α_1 and α_2 domains of the α-chain do indeed sit on top of the rest of the molecule (Plate 7). This structure has been likened to sausages in a pan. The α_1 and α_2 domains each contribute four strands of an eight-stranded β-pleated sheet upon which sit two α-helices formed by other regions of the α_1 and α_2 domains. In between the two helices is a groove that is typical of MHC structure.

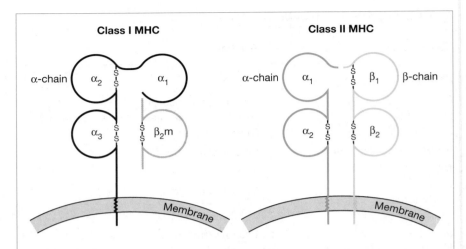

Figure 4.3 Diagram of MHC class I and class II molecules. Class I MHC consists of a 45 kDa α-chain non-covalently attached to β_2-microglobulin. The α-chain of class I MHC has three domains called α1, α2 and α3. Class II MHC consists of a 30–34 kDa α-chain non-covalently linked to a 26–30 kDa β-chain. Both the α- and β-chains of class II contain domains (α_1 and α_2 and β_1 and β_2 respectively). The domains of the class I α-chain, the class II α- and β-chains and β_2-microglobulin are all structurally related to the domains of the Ig and TcR molecules. Note that the α1 and α2 domains of the class I MHC α-chain are not the same as the α_1 and α_2 domains of the class II MHC α-chain.

There are three class I loci in the human, called *HLA-A*, *HLA-B* and *HLA-C*. Each locus has one gene coding for an α-chain, which pairs with β_2-microglobulin to give a complete class I protein.

A second family of class I MHC genes

When most people refer to the class I MHC they are talking about the three genes or proteins called *HLA-A*, *HLA-B* and *HLA-C*. However, there are other class I-like genes and products in the MHC. Some of these other MHC molecules, and their properties are shown in Box 4.3. Their cellular expression and function are not well understood. For the purpose of this book, class I MHC refers to *HLA-A*, *-B* or *-C* or the equivalent in other species.

4.3.3 Class II MHC

Class II MHC molecules also consist of two glycoproteins, an α-chain of molecular size 30–35 kDa and a β-chain of 26–30 kDa. However, unlike class I MHC, they are much more restricted in their expression, being found primarily on cells of the immune system such as monocytes/macrophages and B cells. Both the α- and β-chains are coded for by genes in the MHC. Unlike class I MHC, both the α- and β-chains of the class II MHC molecule have transmembrane regions and cytoplasmic tails (Figure 4.3). Class II MHC molecules are also capable of delivering intracellular signals, another property distinguishing them from class I MHC molecules.

Discovery of a second family of class I MHC molecules

The class I MHC loci in mice are called *H-2D*, *H-2K* and *H-2L*. Different alleles were identified by antibodies raised by immunising mice of one strain with leucocytes or a skin graft from another strain. The mice generated antibodies against the class I molecules of the donor mice that were foreign to the recipient. However, it was observed in some cases that when mice of one strain were immunised with cells from mice that were identical at *H-2D*, *H-2K* and *H-2L*, they made antibodies against cells. Later it was discovered that some of the antigens recognised by these antibodies had a similar structure to class I MHC molecules; they had a long peptide chain with molecular mass of 40–45 kDa associated with β2-microglobulin. However, these molecules differed from *H-2K*, *D* and *L* in two important ways: they were much less polymorphic and their tissue distribution was much more restricted.

Class IB MHC genes and function

When the MHC was examined at the genetic level it was found that there are about 30 genes coding for class I MHC-like molecules, although many of them are pseudo-genes and are not expressed. Similar class I-like genes have been identified in the human; they have been called *HLA-E*, *HLA-F*, *HLA-G*, *HLA-H*, *HLA-J* and *HLA-X*. Because they were discovered after *HLA-A*, *-B* and *-C* they have been referred to as class IB or non-classical class I MHC products. *HLA-A*, *-B* and *-C* are consequently sometimes referred to as class IA or classical class I MHC molecules.

The precise cellular expression and function of the class IB MHC proteins are less well understood. One of the class IB molecules in the mouse has been shown to bind peptides containing *N*-formyl-methionine. These are produced by bacteria but in eukaryotes the *N*-formyl-methionine is removed. Therefore this MHC molecule could specifically bind and present bacterial antigens. Another suggested function of class 1B MHC molecules is protection of the body's own cells from killing by natural killer cells.

Both the α- and β-chains have two external domains with homology to the domains of Ig and class I MHC. The domains furthest away from the cell membrane are the α_1 and β_1 domains of the α- and β-chains respectively and those nearest the cell membrane are called α_2 and β_2. The chains of the class II MHC molecule fold to form a structure that is remarkably similar to class I MHC (Plate 7). The α_1 and β_1 domains form the β-pleated sheet with the two α-helices on top forming the typical 'sausages in the pan' structure seen with class I MHC.

There are three class II MHC loci in the human MHC, called *DP*, *DQ* and *DR* (Figure 4.2). The *DP* and *DQ* loci each contain a functional α-gene, called *DPA* and *DQA* respectively, which codes for the α-chain, and a functional β-gene, called *DPB* and *DQB*, which codes for the β-chain. The *DR* locus is more complicated; it contains one functional α-gene, called *DRA*, but different versions of the locus contain one or two functional *DRB* genes, each coding for a separate β-chain (see Figure 4.2).

4.3.4 Polymorphism of class I and class II MHC

One of the extraordinary features of class I and class II MHC genes is the degree of polymorphism shown at the different loci. A particular gene can exist in slightly different forms in different individuals; these different forms are called alleles. Alleles arise when the gene for a protein mutates to give a different DNA sequence, which may result in a different amino acid sequence in the protein. Many of these mutations are lost by chance or because the mutated protein does not function as well as the original. However, in some cases the different forms of the gene become established in the population as alleles. Where a gene expresses different alleles it is said to be polymorphic.

Many genes are not polymorphic and only one version of the protein exists; these genes are said to be monomorphic. Where genes are polymorphic they usually have only a few alleles (fewer than five) and the different forms of the protein differ by one or a few amino acids. Class I and II MHC genes are unique both in the number of alleles at the different loci and also by the degree to which the amino acid sequence of the alleles can vary. Some MHC loci in the human have over 100 alleles and the amino acid sequence between alleles can vary by more than 20 amino acids (see Table 4.2). Therefore the variety of different MHC molecules that can be expressed in a population vastly exceeds the variety of any other protein. The possible reason for this extraordinary polymorphism will be explained when the function of MHC molecules has been described.

4.3.5 Nomenclature of MHC genes, proteins and alleles

The extraordinary polymorphism of the class I and class II MHC loci led to a predicted confusion in terminology. There is now a standard nomenclature for the alleles of the different MHC genes, which ensures consistency but is not that easy to understand unless a little is known about its development.

Table 4.2 Number of alleles of class I and class II MHC genes

Gene	Number of alleles
HLA-A	323
HLA-B	727
HLA-C	209
HLA-DPA	22
HLA-DPB	119
HLA-DQA	31
HLA-DQB	67
HLA-DRA	3
HLA-DRB	501[a]

These are the number of DNA sequences at each locus and do not necessarily represent different protein structures.

[a]Total number or alleles for all *DRB* loci.

Initially the different alleles of the MHC were recognised using antibodies. Individuals who are exposed to foreign MHC molecules can make antibodies against them just as they would against any foreign antigen. Therefore women who have had babies (especially more than one), people who have had blood transfusions and some people who have had transplants have all been exposed to foreign MHC molecules and their sera may contain antibodies that are specific for the foreign MHC molecule. Using these sera it is possible to identify specific MHC molecules on the surface of leucocytes so that the particular alleles of MHC an individual possesses can be established. This process is known as **tissue typing** (Figure 4.4).

These sera showed that there were many alleles at *HLA-A, -B, -C* and *–DR* (they were not so useful for *DP* and *DQ*) and many different forms of the proteins. The different alleles of a particular MHC molecule that were identified by antibodies were given different numbers. Therefore different alleles of *HLA-A* were called *HLA-A1, HLA-A2, HLA-A3*, etc. Similarly alleles of *HLA-B* were called *HLA-B1, HLA-B2*, etc. and the same was adopted for *HLA-C* and *DR*. These were the **serologically determined** alleles.

With the advent of DNA sequencing and other molecular genetic techniques for examining HLA polymorphism it became apparent that there

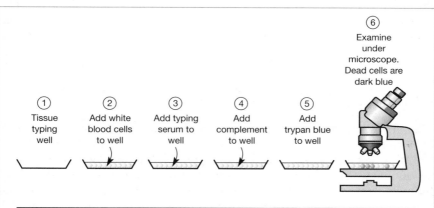

Figure 4.4 HLA-typing. The table shows a simplified reaction pattern of three typing sera with cells of three different *HLA-A* types. A tick indicates that the serum recognises the HLA type of the cells and kills them. A cross indicates that the serum does not recognise the HLA on the cells and does not kill them.

were many more HLA alleles than could be identified by antibodies. For example *HLA-A2* was one serologically determined allele of *HLA-A*. However, when people who were serologically *HLA-A2* were typed genetically (including sequencing the DNA of the *HLA-A2* gene in different individuals) it was found that there were 13 different genetic variations of the serologically determined *HLA-A2*. Because these genetic alleles are similar to each other, the antibodies recognising *HLA-A2* could not distinguish between them and serologically they all appeared the same. The nomenclature of HLA was therefore amended to take into account both the serological and genetic bases of the polymorphism.

As described in Section 4.3.2, the genes for class I MHC are called *HLA-A*, *HLA-B* and *HLA-C*. The genes for class II MHC α-chains are *DPA*, *DQA* and *DRA*. The genes for the class II MHC β-chains are *DPB*, *DPQ* and *DRB1*, *DRB3*, *DRB4* and *DRB5*; there are four different genes coding for the DR β-chain and an individual will have either *DRB1* alone or *DRB1* and *DRB3*, *DRB4* or *DRB5*. The alleles for each gene are given a four-number designation. The first two numbers refer to the serologically determined allele and the last two numbers refer to the genetically determined allele. Continuing with *HLA-A2* as an example, the serological allele is now called *HLA*A02*. The 13 different genetic alleles have been assigned the numbers 01 to 13. The full description of the alleles is now *HLA-A*0201*, *HLA-A*0202*, etc. to *HLA-A*0213*. The same nomenclature was adopted for the other MHC genes and Table 4.2 shows the number of different alleles of the class I and class II MHC genes.

4.3.6 How many different class I and class II MHC proteins does an individual express?

Like any other gene, individuals inherit one copy of an MHC gene from their mother and one from their father. Given the extreme polymorphism of the MHC genes it is likely that an individual will inherit two different alleles of a gene. Such an individual is said to be heterozygous for that gene. Individuals who inherit two copies of the same allele are homozygous for that gene. MHC genes are expressed co-dominantly; that is, if you inherit two different alleles of a gene you will make both forms of the protein. Therefore individuals can express many different MHC molecules but may vary in exactly how many. The different MHC molecules an individual can express are now described.

Class I MHC
The situation with class I is relatively simple. An individual can express one (if they are homozygous) or two proteins of each of HLA-A, -B and -C. Therefore an individual will express between three and six class I MHC proteins.

Class II MHC
This is more complicated because both the α- and β-chains are polymorphic. When the two class II MHC chains associate to form class II molecules, the

α-chain from one locus can associate only with a β-chain from the same locus. Therefore DPα can associate with DPβ but not with DQβ and DQα can associate only with DQβ. The DRα can associate with any of the DRβ chains expressed by an individual.

However, each α-chain for a locus can associate with more than one β-chain to generate more than one class II MHC product for each locus. To take DQ as an example, you will inherit a maternal *DQA* gene and a paternal *DQA* gene as well as a maternal and a paternal *DQB* gene. The α-chain coded by the maternal *DQA* gene can associate with the β-chain coded for by the maternal *DQB* gene or the paternal *DQB* gene (Figure 4.5). Similarly the paternal α-chain can associate with the paternal or the maternal β-chain. If you are heterozygous for both the α and β genes, you can make four different DQ molecules. If you are homozygous at both α and β, you can make only one DQ molecule. The situation is exactly the same with DP: you can make up to four different DP molecules if you are heterozygous at *DPA* and *DPB*.

The situation with DR is simpler in some ways but more complicated in others. The *HLA-DRA* gene is not functionally polymorphic and therefore there is only one form of the DRα-chain. However, an individual may make between one and four different DRβ-chains depending on which alleles they inherit (see Figure 4.2) and whether they are heterozygous. Therefore it is possible to express between one and four HLA-DR molecules.

Overall then the fewest different class II MHC molecules you could make would be three (one each of DP, DQ and DR) and the most would be 12 (4DP + 4DQ + 4DR). Most individuals express a number between these two extremes.

Figure 4.5 *Cis*- and *trans*-association of class II MHC α- and β-chains. The example uses *HLA-DQ* and assumes the individual is heterozygous for *DQα* and *DQβ*. An individual inherits one set of *DQ* genes from their father and one set from their mother. Paternally inherited genes and their products are shown in dark blue and maternally inherited in light blue. Because MHC genes are expressed codominantly, two different DQα chains, α_p and α_m, and two different DQβ chains, β_p and β_m, are made. Each α-chain can associate with either β-chain. The association of α_p with β_p and the association of α_m with β_m are called *cis*-association because the genes coding the α-chain and the β-chain are on the same chromosome ①. By contrast, association of α_p with β_m or α_m with β_p is called *trans*-association because the genes coding the α- and β-chains are on different chromosomes ②.

4.4 Recognition of antigen by T cells

The similarity in structure between class I MHC and class II MHC molecules and the unique level of polymorphism in the class I and class II MHC genes suggest that class I and class II MHC perform a similar function. In fact they do, and this function is to help T cells recognise antigen. Both class I and class II MHC molecules have a groove on top of the molecule (Figure 4.3 and Plate 7). Small peptides derived from antigens bind to the groove of the MHC molecules so that MHC molecules on the surface of a cell display these antigenic peptides (Figures 4.3 and 4.6 and Plate 7). The complexes of peptide/MHC are sticking out from the cell surface so that they are easily accessible to the TcR on a T cell. The TcR can bind to the complex of the antigenic peptide and the MHC molecule; some regions of the TcR antigen-binding site bind to parts of the antigenic peptide and some bind to parts of the α-helices of the MHC molecule (Figure 4.6 and Plate 8). There is one very important difference in the way CD4 and CD8 T cells recognise antigen. CD4 T cells recognise antigen presented by class II MHC molecules and CD8 T cells recognise antigen presented by class I MHC molecules.

The binding of the TcR to the antigen peptide/MHC complex is specific for both antigen and MHC. The variable regions of the α- and β-chains of the TcR fold to form the antigen/MHC binding site. Both CD4 and CD8 T cells

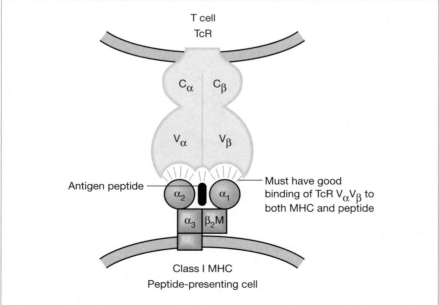

Figure 4.6 Interaction of the TcR with antigen/class I MHC. The antigen-binding site of the TcR is made up of parts of the Vα and Vβ chains. Parts of the binding site interact with the antigenic peptide bound in the groove of the MHC molecule and parts of the binding site interact with the MHC molecule.

use the same TcR genes to generate their TcRs but, just like antibody, each T cell will have different amino acid sequences in the hyper-variable regions of the α- and β-chains of the TcR and therefore will have a different shape in the antigen/MHC binding site of the TcR. The binding of the TcR to the antigen/MHC complex involves the same types of non-covalent bonds as antibody binding to antigen (see Section 3.3.2). Only if the TcR fits closely to *both* the antigenic peptide and the MHC molecule will binding occur. Therefore different T cells will have different specificities for antigen/MHC.

An individual T cell will recognise antigen only in association with one MHC molecule. Therefore the same antigen being presented by different MHC molecules will be recognised by different T cells (Figure 4.7). For instance a CD8 T cell that can recognise an antigenic peptide presented by an *HLA-A* molecule will not be able to recognise the same peptide presented by an *HLA-B* molecule. In fact it will not even be able to recognise a peptide presented by a different allele of an HLA-A, e.g. one TcR will recognise antigen presented by *HLA-A1* but not the same antigen presented by *HLA-A2*. This is what is meant by **MHC-restricted recognition of antigen** that is seen with T cells but not B cells.

4.5 Antigen processing and presentation by MHC molecules

We have seen how the role of class I and class II MHC molecules is to bind peptides derived from antigen and to present these peptides for recognition by the TcR on T lymphocytes. These peptides are 7–18 amino acids in length and are derived from antigenic proteins that can be up to 1000 amino acids in size. The question arises: how does the antigenic peptide get to the MHC molecule? The answer is that large protein antigens are broken down into peptides inside cells and these peptides associate with MHC molecules intracellularly. The MHC molecules bearing the antigenic peptides are then transported and expressed on the cell surface. However, the way in which this occurs is quite different for class I and class II MHC molecules and is also influenced by where the original antigen is located. Endogenous antigens are produced within the cell (e.g. viral proteins) and are processed and presented by class I MHC. Exogenous antigens derive from outside the cell (e.g. from an extracellularly living bacterium or parasitic worm) and are processed and presented by class II MHC.

4.5.1 Presentation of endogenous antigen by class I MHC

When a virus infects a cell its DNA directs the production of viral proteins within the cytoplasm of the cell (Figure 4.8). The object of this exercise from the virus's point of view is to generate new viral proteins that eventually will be used to generate new viral particles. Because the viral proteins are in the cytoplasm of the cell they are inaccessible to antibody. However, peptides derived from these viral proteins can be presented by class I MHC. This occurs in the following way.

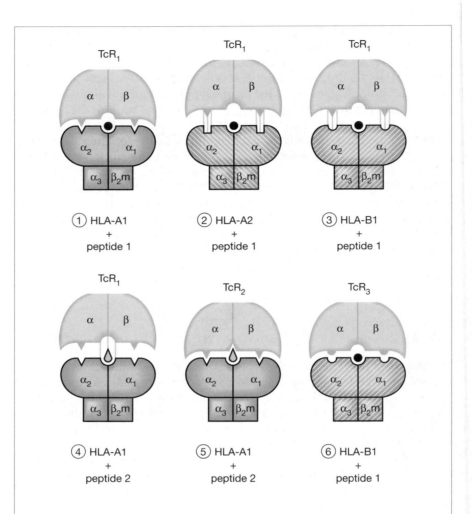

Figure 4.7 MHC-restricted recognition of antigen by T cells. TcR$_1$ can recognise antigenic peptide 1 being presented by HLA-A1 ①. TcR$_1$ cannot recognise antigenic peptide 1 presented by HLA-A2 ② or HLA-B1 ③; nor can it recognise antigenic peptide 2 being presented by HLA-A1 ④. TcR$_1$ is therefore said to be specific for antigenic peptide 1 restricted to HLA-A1. TcR$_2$ can recognise a different antigenic peptide (peptide 2) being presented by the same HLA-A1 ⑤ as TcR$_1$ and is therefore specific for antigenic peptide 2 restricted to HLA-A1. Finally TcR$_3$ can recognise peptide 1 in association with HLA-B1. Antigenic peptides are shown in black or grey.

Proteosomes degrade cytoplasmic antigens

Within the cytoplasm of cells are structures called proteosomes (Figure 4.8). Proteosomes are made up of 12–15 protein subunits, which assemble to form a tube-like structure. At least two of the proteins, called LMP-2 and LMP-7, that make up proteosomes are coded for by genes that are found in the class II region of the MHC. As their name implies, proteosomes have proteolytic activity and are able to degrade cytosolic proteins into peptide

Figure 4.8 The class I MHC pathway of antigen processing. Endogenous antigen (produced within the cell) is degraded in proteosomes and antigenic peptides are transported by TAP to the RER where they associate with the class I molecule being assembled. The class I molecule bearing the antigenic peptide is then transported to the cell surface and presents the antigenic peptide. A protein, p88, is initially associated with class I in the RER but dissociates when antigen peptide is bound.

fragments of 5–15 amino acids. The optimal size of peptide fragments for binding to class I MHC is 8–10 amino acids.

Transporters carry antigenic peptides into the rough endoplasmic reticulum
The next problem is how to get the peptides generated by proteosomes to bind to class I MHC molecules. Like any other cell surface protein, class I MHC molecules are synthesised in the rough endoplasmic reticulum (RER), modified in the Golgi apparatus and transported to the cell surface in transport vesicles. The peptides and MHC molecules are in separate compartments of the cell. The situation is resolved by the presence of special peptide transporter molecules that transport the peptide fragments from the cytoplasm to the RER. These transporters are made up of two proteins called TAP-1 and TAP-2 (Figure 4.8). Like the genes for LMP-2 and LMP-7, the genes for TAP-1 and TAP-2 are located in the class II region of the MHC. TAP stands for **t**ransporter associated with **a**ntigen **p**rocessing. The transporter composed of TAP-1 and TAP-2 is a member of a family of transporter molecules called ABC transporters. These have ATP-binding cassettes (hence the term ABC) and use the energy from the hydrolysis of ATP to move ions or small proteins across membranes. The TAP transporter is most efficient at transporting peptides of 8–15 amino acids in length.

Antigenic peptides are required for the correct assembly of class I MHC molecules
Once the antigen-derived peptides have been transported to the RER by the TAP transporters they need to bind in the groove of the class I MHC mole-

cules. This process is more complex than simple binding of the peptide to the groove of a completed class I MHC molecule. The proper association of β_2-microglobulin with the class I MHC α-chain to form a stable class I MHC molecule requires the presence of an antigenic peptide. Thus the whole complex of class I MHC α-chain, β_2-microglobulin and antigenic peptide is assembled. The class I MHC with the antigenic peptide in its groove then travels to the Golgi and subsequently is transported to the cell surface in a transport vehicle and is expressed on the cell surface where it can present the antigenic peptide to CD8 T cells. In the absence of antigenic peptide the class I MHC molecules formed in the RER are unstable and do not travel to the cell surface. This process ensures that only class I MHC molecules bearing antigenic peptide, and hence able to present the peptide to T cells, are expressed on the cell surface and empty class I molecules, which serve no useful purpose, do not clutter up the cell surface.

4.5.2 Presentation of exogenous antigen by class II MHC

Class II MHC molecules present antigenic peptides that are derived from antigens that are outside the cell. The fundamental problems are the same as with class I MHC – how do you generate antigenic fragments from larger proteins and get them to bind to class II MHC? The pathway for this is quite different from that for class I MHC.

Extracellular proteins are endocytosed and degraded in lysosomes
Extracellular proteins, which could be antigens derived from an extracellular dwelling pathogen, are taken up by cells by endocytosis. This could be pinocytosis, receptor-mediated endocytosis or phagocytosis (Figure 4.9). The endocytic vesicle, or endosome, containing antigen proteins lowers its pH and fuses with a lysosome to form an endolysosome. The contents of the endosome are now exposed to the proteolytic enzymes of the lysosome. These enzymes are active at low pH and begin to degrade the antigen into smaller fragments.

Antigenic peptides associate with class II MHC in specialised compartments
Class II MHC molecules, like class I, are synthesised in the RER and travel via the Golgi to the cell surface in vesicles. Again the problem is how to get the peptide, which is in one type of vesicle, to bind to class II MHC, which is in another. It has recently been found that this occurs in a special vesicle called the compartment for peptide loading (CPL). When class II MHC is initially synthesised in the RER, the α- and β-chains associate with a third protein chain called the **invariant** chain (Figure 4.9). The invariant chain is thought to have two important functions. It binds to the groove of the class II MHC molecule and therefore stops self-peptides being made in the RER from binding to the MHC. The invariant chain is also thought to direct the class II MHC/invariant chain complex from the Golgi to the CPL. Because class I molecules are not associated with invariant chain they do not go to the CPL. In this way class I and class II MHC molecules, which are both

Figure 4.9 The class II MHC pathway of antigen processing. Exogenous antigen (produced outside the cell) is taken up by endocytosis ①. The endosome fuses with a lysosome ② and the antigen is degraded ③. Class II MHC is initially synthesised in the RER with an associated protein, the invariant chain ④. The endocytic vesicle delivers the antigenic peptides to a compartment for peptide loading ⑤ where the antigenic fragment displaces invariant chain from class II MHC and becomes bound to the MHC molecules ⑥. The MHC molecules are then transported to the cell surface ⑦ where they present antigenic peptide.

present in the same Golgi network, are directed to separate vesicles – class I to a standard transport vesicle and class II to the CPL.

The CPL now fuses with the endolysosome containing the degraded antigenic peptides and, in a way that is not fully understood, antigenic peptides of the right size (5–13 amino acids) displace the invariant chain bound to the class II MHC and become bound in the groove of the MHC molecule. The class II MHC molecule bearing the antigenic peptide is then transported to the cell surface where it is expressed and can present the antigenic peptide to CD4 T cells.

4.5.3 The importance of two antigen-processing pathways

The two pathways of antigen processing ensure that exogenous antigen is targeted to class II MHC for recognition by CD4 T cells and endogenous antigen is targeted to class I MHC for recognition by CD8 T cells. This is very important in relation to the different biological function of CD4 and

CD8 T cells (see Chapters 6 and 9). Both cell types interact with other cells and recognition of antigen/MHC is an important part of this interaction. However, CD8 T cells may be required to interact with almost any cell type and therefore recognise antigen presented by class I MHC, which is expressed on most cell types. CD4 T cells interact mostly with other cells of the immune system that express class II MHC.

4.5.4 The nature of binding of peptides to MHC

The binding of peptides to MHC molecules shows some specificity but not to the same degree as, for instance, the binding of antigen to antibody or the TcR to antigen/MHC. Generally class I peptides will bind to many different class I MHC molecules and alleles, but not all of them. Therefore a peptide may bind to HLA-A1 and HLA-B1 but not HLA-A2 or HLA-B2. Class II peptides are more promiscuous in their MHC binding and generally a class II peptide will bind to a larger variety of class II MHC molecules and alleles than is seen for class I peptides binding to class I MHC.

4.5.5 The relevance of MHC polymorphism

Two questions raised by the arrangement of the MHC complex are: why is the MHC so polymorphic and why do you have three different class I loci doing the same thing and three class II loci doing the same thing? The answers to these questions are related and can best be explained if you imagine a situation where you had only one class I MHC locus and one class II MHC locus and there was no polymorphism. One survival strategy for pathogens is to evolve antigens that produce peptides that cannot bind to MHC molecules in the host; obviously in this situation there is no possibility of T cells recognising the antigen. With only one MHC locus and no polymorphism a pathogen would need only to evolve antigens that did not bind to this one MHC molecule and the whole population would be at risk from infection and possible elimination (Figure 4.10). If an individual has three loci for each type of MHC molecule and therefore expresses three different MHC molecules, it is much more difficult for a pathogen to evolve antigens that do not bind to any of the three different MHC molecules. If you now add extreme polymorphism it means that the population as a whole expresses many different variants of MHC molecules. This makes it almost impossible for a pathogen to evolve antigens that would not bind to at least some of the MHC variants. Therefore the population as a whole is protected, even though some individuals may not respond to pathogen antigens. Evidence is beginning to emerge that MHC polymorphism can be driven by selection of resistance to endemic infectious agents. In parts of Africa where malaria is common, the frequency of certain HLA alleles is higher than in other populations where malaria does not occur. These alleles have been shown to be associated with increased resistance to the parasite.

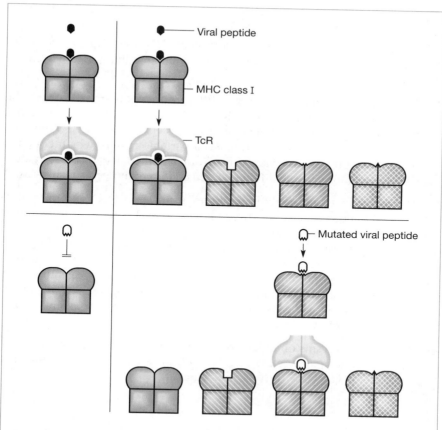

Figure 4.10 The advantage of multiple MHC loci and MHC polymorphism. The left-hand panel illustrates the situation if we possessed only one class I HLA locus that was not polymorphic. Although a viral peptide may bind to the HLA molecule (top panel), if the gene for the peptide mutated the antigenic peptide may no longer bind to the HLA and therefore no recognition of the viral peptide would be possible and the individual would be susceptible to disease caused by the virus. In the right-hand panel it can be seen that if an individual has multiple HLA class molecules, each with a different capacity to bind antigenic peptide, a similar mutation in the pathogen may result in an antigenic peptide that does not bind to the original HLA molecule but the peptide will be bound and presented by a different HLA class I molecule, thereby maintaining the ability to respond to the virus.

4.6 Summary

- There are two types of T lymphocytes called CD4 T cells and CD8 T cells. Both are produced in the thymus from precursors that migrate from the bone marrow.

- T cells recognise antigen using a different receptor from B cells. The T cell receptor for antigen (TcR) has two chains, an α-chain and a β-chain. It is found in the membrane of T cells but is not secreted like antibody. Each chain of the TcR has a variable region, which contains hyper-variable regions that differ from T cell to T cell; these hyper-variable regions give the different T cells different antigen specificities.

- The major histocompatibility complex is a region of DNA containing over 100 genes. There are three classes of MHC genes and products called class I, class II and class III MHC genes. Class I and class II MHC genes are extremely polymorphic, with some genes having over 100 alleles. Class I and class II MHC proteins are expressed on the surface of cells and present antigenic peptides to the TcR. T cells therefore recognise a complex of antigen + MHC. CD4 T cells recognise antigen presented by class II MHC molecules and CD8 T cells recognise antigen + class I MHC.

- Special mechanisms of antigen processing ensure that endogenous antigens are presented by class I MHC molecules and exogenous antigens are presented by class II MHC molecules.

CHAPTER FIVE

Generation of diversity:
how do T and B cells generate so many different variants of their antigen receptors?

Learning objectives

To know the organisation of Ig and TcR genes. To understand the process of Ig and TcR gene rearrangement. To understand how unique molecular mechanisms enable hundreds of genes to generate millions of different variants of Ig and TcR molecules.

Key topics

- Structure of the genes coding Ig and TcR
- Gene rearrangement
- Generation of variety in Ig and TcR V regions
  - Selection of gene segments
  - Junctional diversity
  - Somatic mutation
  - Association of Ig H+L or TcR α+β-chains
  - Number of Ig and TcRs

5.1 Introduction

The immune system is faced with a considerable problem in having to generate a huge number of different antigen receptors with specificity for a seemingly endless variety of antigens. The number of different pathogens in the world is vast and each pathogen will have its own set of antigens made up of proteins, carbohydrates or lipids. Although it is impossible to know how many antigenic epitopes there are, a reasonable estimate is that in excess of 10^{11} different antibody specificities are required to recognise all the potential antigenic epitopes (see Box 5.1 for an explanation). Different antibodies will be needed to recognise all these different antigens. The number of T cell specificities required is similarly large.

BOX 5.1: HOW MANY ANTIGENIC EPITOPES ARE THERE?

It is impossible to know how many epitopes there are. It is possible to calculate an estimate of the number of *potential* epitopes. X-ray crystallography of antibody–antigen interactions has revealed that an antigenic epitope comprises 8–22 amino acids depending on the nature of the antigen. There are 20 different amino acids. Therefore, in a peptide, each amino acid position has 20 alternatives. With a peptide of one amino acid there are 20 possible alternatives (the number of amino acids) $= 20^1$. With a two-amino-acid peptide (dipeptide) there are many more different forms possible. If the first amino acid is, say, alanine, it is possible to combine this with any of the 20 amino acids at position 2 giving rise to 20 alternative dipeptides. Changing the first amino acid to arginine enables this to be combined with any of the 20 amino acids at the second position, giving another 20 dipeptides. This process can be repeated for each of the 20 amino acids at position 1; each of the 20 amino acids at position 1 gives rise to 20 dipeptides by being combined with a different amino acid at position 2. Therefore the number of combinations is 20×20, or 20^2. You can see that the number of combinations is determined by the number of different amino acids, 20, and the number of amino acids in the peptide, n, i.e. 20^n. For one amino acid, where $n = 1$, the number of possibilities is $20^1 = 20$. For two amino acids the number of combinations is $20^2 = 400$.

If we take the lower value of eight amino acids for an antigenic epitope, there are a total of $20^8 = 2.56 \times 10^{10}$ combinations. The number of combinations of 22 amino acids is $20^{22} = 2 \times 10^{28}$. It is probable that only a tiny fraction of these combinations exists but even if one in a million of these combinations existed, the total number of epitopes would be in excess of 10^{22}.

Although the exact number is not known it can be seen that the potential number of epitopes is very large indeed and a reasonable estimate is in excess of 10^{11}. Therefore the immune system needs to be able to generate a large number of different antigen specificities to be able to recognise this large number of epitopes.

No individual will ever encounter all the epitopes that exist, even over a lifetime. However, for the immune system this does not matter. The specific immune system must cater for all *possibilities*. The immune system cannot know which epitopes exist and, more importantly, cannot know which epitopes an individual may be exposed to. Therefore the immune system must generate B and T cells with specificities for all the potential epitopes that an individual may encounter. So how can the immune system generate enough different Ig and TcR molecules with different antigen specificities to recognise over 10^{11} different epitopes? It is not possible to devote a separate gene for each Ig or TcR specificity. Given that the human genome contains about 30 000 genes coding for proteins, even if every gene was devoted to making Ig or TcR there would not be enough genes to generate anything like the variety of Igs and TcRs to recognise more than 10^{11} potential epitopes. To solve the problem of generating huge numbers of different receptor specificities from a limited amount of genetic material, the immune system uses unique molecular mechanisms involving rearrangement of the genes coding for the Ig and TcRs.

5.2 The relationship between Ig and TcR genes and the proteins produced

Antibody molecules consist of two heavy chains and two light chains joined together by disulphide bonds (see Chapter 3). The TcR consists of an α-chain and a β-chain (see Chapter 4). The locations of the Ig and TcR genes are shown in Table 5.1. Each Ig or TcR chain has a variable (V) region and a constant (C) region. However, there are unique features about the genes that can best be explained by describing the human Ig κ-light chain gene, which has the simplest structure. The details of the other genes will be described later in the chapter.

5.2.1 Gene and protein structure of the human κ-light chain

At the protein level the κ-light chain consists of a variable region domain and a constant region domain. The genes coding for the κ-light chain consist of exons, which are sequences of DNA coding for the protein, and introns, which are parts of the gene that do not code for the protein. In this respect the κ-light chain gene is similar to most other eukaryotic genes. Figure 5.1 shows the structure of a typical gene, that for α-globin, as well as the κ-light chain gene. Superficially they appear to be similar in that they both contain introns and exons. The α-globin gene has been chosen for its simplicity: it contains three exons and two introns. Other genes can contain dozens of introns and exons. The primary and mRNA transcripts for α-globin are shown in Figure 5.1 and it can be seen that all of the exons are used to code for the final protein sequence. This is typical for most genes although sometimes the primary RNA can be processed in slightly different ways – so-called differential processing – to give different forms of the protein. However, this differential splicing results in only a few different forms of the protein.

If we now look at the κ-light chain gene, two important features are seen that distinguish this gene from typical genes. At the protein level the κ-chain consists of a variable and a constant region. A single exon codes for the constant region of the κ-chain; this is often referred to as the 'Cκ gene' although strictly speaking it is a gene segment. The variable region is coded by two exons. Most of the variable region is coded for by a single exon

Table 5.1 Location of Ig and TcR genes

Gene[a]	Chromosome
IgH	14
κ	2
λ	22
TcRα	14
TcRβ	7

[a] The TcRδ genes are located between those coding for Vα and J.

called the κ variable gene or Vκ (Figure 5.1). However, the part of the variable region closest to the constant κ region is coded for by a separate gene segment called the J gene or Jκ. The J stands for 'joining' because the J segment joins the V and C segments. The unique property that the gene shares with the other Ig and TcR genes is that there are a large number of V genes (or exons) and, to a lesser extent, J genes but *each B cell uses only one of the V genes and one of the J genes*. As shown in Figure 5.1, the DNA of the gene in a B cell has one of its 76 V genes joined to one of its five J genes and it is from this VJ DNA sequence that RNA will be transcribed. This 'choosing' to utilise only one out of the many exons occurs because of gene rearrangement during the development of the B cell (see Section 5.3) and is unique to the Ig and TcR molecules. Although each individual B cell uses only one V and one J segment, different B cells will 'choose' different V and J segments at random. Therefore the population of B cells in an individual will utilise all the different V and J segments (Figure 5.2).

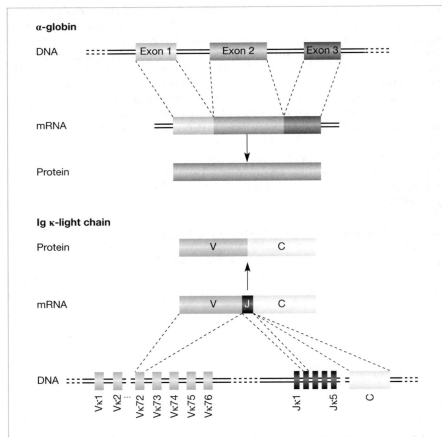

Figure 5.1 Structure of genes coding for α-globin and κ-light chain. All three exons of the α-globin gene are translated into RNA, which is processed and translated into protein. By contrast, only one of 76 Vκ genes and one of five Jκ genes are transcribed into RNA, along with the Cκ gene.

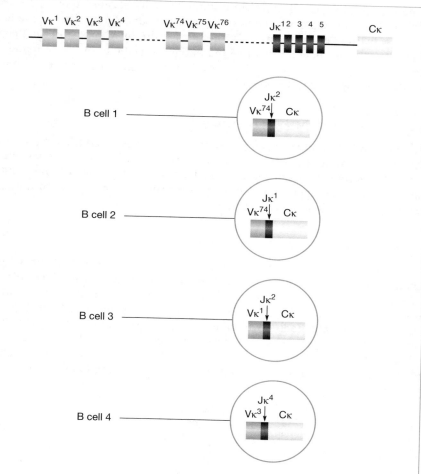

Figure 5.2 Different B cells utilise different Ig gene segments. Using the Ig κ-light chain as an example, there are 76 genes coding for Vκ (42 of these are pseudo-genes and are not used) and five genes coding for Jκ. Each B cell chooses at random one functional Vκ gene and one Jκ gene; the actual genes chosen will differ for each B cell. Therefore B cell 1 has chosen the Vκ74 and Jκ2. B cell 2 has chosen the same V, Vκ74, but a different J, Jκ1, and it can be seen that all four B cells have a different combination of V and J regions. Each V and J region has a unique nucleotide sequence and therefore each combination of V and J gives rise to a V-region protein with a slightly different amino acid sequence, and therefore different antigen specificity, from any other combination of V and J.

An important feature of the gene organisation of the κ gene is that the J gene codes for part of the third complementarity-determining region (CDR3) of the variable region and is therefore important in antigen specificity. It was discussed in Chapter 3 how Ig chains contain three regions that differ extensively in different antibodies. These regions contribute to the antigen-binding site. CDR1 and CDR2 are coded for exclusively by the

V gene segment but will differ in different B cells because they will select different V gene segments. CDR3 is coded for partially by the V gene and partially by the J gene (Figure 5.3); therefore both the V gene and the J gene that a B cell selects will influence the antigen specificity of the Ig made by the B cell. This feature is very important in increasing the variety of amino acid sequences in CDR3 (see Section 5.3.1).

5.2.2 Structure of the λ-light chain gene

The human λ-light chain gene is slightly more complicated than the κ gene. There are 30 functional Vλ genes but instead of a series of J genes and a single C region gene as in the κ gene, there are four repeats of functional J and C genes (see Figure 5.4). The λ chains produced from each of the four Cλ genes are structurally the same.

5.2.3 Structure of Ig heavy chain gene

The basic organisation of the Ig H genes is similar to that of the κ-light chain genes but there are some important differences (Figure 5.4). Instead of the variable region being coded for by V and J genes there is an additional set of exons, called D, between the V and J genes. The D in this case stands for diversity. Each developing B cell 'chooses' one D gene as well as one V and one J gene to use in generating the H chain. The D gene codes for part of the CDR3 of the heavy chain, so that three genes – part of V, and all of D and J – contribute to coding for the CDR3 of the Ig heavy chain.

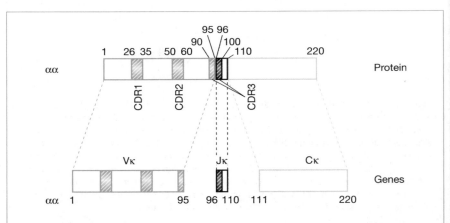

Figure 5.3 Relationship between Ig κ gene and protein. The Vκ-gene codes for the first 95 amino acids of V region of the protein. The Jκ gene codes for the final 15 amino acids of the V region of the protein. The CDR1 and 2 are coded for totally by the Vκ gene but the CDR3 is coded for partially by the Vκ gene and partially by the Jκ gene. Therefore different combinations of V and J will give different nucleotide sequences coding for CDR3 and therefore different amino acid sequences at CDR3. Because CDR1, 2 and 3 fold in the antibody molecule to form the antigen-binding site, the different amino acid sequences at CDR3, formed by different combinations of Vκ and Jκ genes, will have different antigen specificities.

Figure 5.4 Organisation of immunoglobulin and TcR genes. The basic organisation of all the genes is similar with V-region genes, J-region genes, C-region genes and, in the case of IgH and TcRβ, D-region genes. Note that the C region for IgH actually consists of seven exons and the C region for TcRβ consists of four exons.

Downstream (3') of the J genes are the constant region genes, which code for the constant region part of the molecule. There is one Ig heavy chain constant region gene for each class (and subclass) of antibody, giving a total of eight constant genes in humans. These are called Cμ, Cδ, Cγ1, Cγ2, etc. for the different Ig classes. Each constant region gene has seven exons (Figure 5.4).

5.2.4 Structure of the TcR genes

The TcR genes are similar in arrangement to the Ig genes (Figure 5.4). The TcR α gene consists of one C region, which is coded for by four exons. There are about 50 Vα genes and a surprisingly high number of J genes – also 50 – in the α gene. Just as with the Ig light chain genes, each T cell 'chooses' one V gene and one J gene to use for its TcR α-chain. The CDR3 is coded for by part of the V and part of the J gene, while the CDR1 and CDR2 are coded for by the V gene.

The TcR β gene is similar to the Ig heavy chain gene in that it utilises V, D and J segments to code for the variable part of the molecule. The TcR β gene consist of about 50 V genes, followed by two repeats of a cluster of one D-region gene, seven J-region genes and a constant region gene coded for by four exons. There are a total of two D genes, 13 J genes (one J gene is a pseudo-gene and not expressed) and two constant region genes in the TcR β gene.

Unlike the Ig heavy chains, whose constant regions give them different functional properties, the two constant regions in the β-chain do not have any different functions. Both CD4 and CD8 T cells use either constant region randomly and it would appear that the TcR is used just to recognise antigen. There is no secreted TcR equivalent of secreted Ig.

5.3 Rearrangement of receptor genes in B and T cells

So far we have talked about B and T cells 'choosing' one out of many V, D or J segments to use for coding for their Ig or TcR proteins. How do B or T cells 'choose' which segments to use? Obviously the cells do not choose which genes to use in the sense that we would choose which brand of baked beans to buy in the supermarket. During their development, B and T lymphocytes differentiate from multipotential stem cells in the bone marrow into specialised B and T cells. As part of this process, developing B cells will rearrange an Ig heavy chain gene and one light chain gene, either κ or λ, and developing T cells will rearrange their TcR α and β genes.

Gene rearrangement involves taking different gene segments and 'sticking' them together. Using the Ig heavy chain gene to illustrate the concepts, the V genes are located 5' of the D genes, which are themselves 5' of the J genes. During gene rearrangement one V gene moves and its 3' end is 'stuck' to the 5' end of one of the D genes. The 3' end of that D gene also moves and is 'stuck' to the 5' end of one of the J genes (see Figure 5.5). At the molecular level these rearrangements involve specific recognition sequences and enzymes that bring about gene rearrangement. The recognition sequences for gene rearrangement are special DNA sequences that occur 3' and 5' of rearranging genes (see Box 5.2).

The particular V, J and D genes chosen by an individual cell are random. One developing B cell will choose one particular V_H gene, one D_H gene and one J_H gene to rearrange for its Ig H-chain, while a second B cell will choose different V_H, D_H and J_H genes. The same randomness of choice occurs for light chain V and J segments in B cells and TcR V, D and J gene segments in T cells. As described for the κ gene in Section 5.2.1, because different lymphocytes choose different gene segments, all the gene segments are used by the populations of lymphocytes in an individual.

5.3.1 Having separate V, D and J genes greatly increases Ig and TcR diversity

Having V, D and J genes coding for CDR3 increases the number of varia-tions of CDR3s considerably. Since CDR3 contributes to antigen binding, an increased number of CDR3s means an increased number of antigen specificities. Using Ig as an example, there are 51 V_H genes and 70 V_L genes

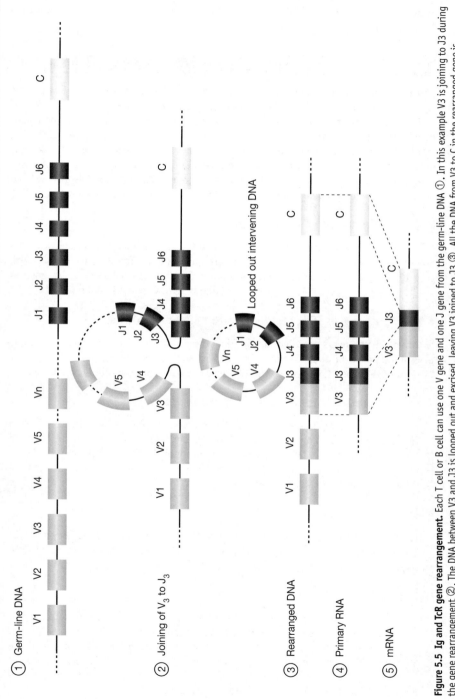

Figure 5.5 Ig and TcR gene rearrangement. Each T cell or B cell can use one V gene and one J gene from the germ-line DNA ①. In this example V3 is joining to J3 during the gene rearrangement ②. The DNA between V3 and J3 is looped out and excised, leaving V3 joined to J3 ③. All the DNA from V3 to C in the rearranged gene is transcribed to give the primary transcript ④ and this is processed to give the mRNA ⑤, which is translated into protein.

$(40\kappa + 30\lambda)$. Since any heavy chain can pair with any light chain, there would be $51 \times 70 = 3570$ different combinations of $V_H + V_L$ if we did not have separate V, D and J genes.

BOX 5.2: Ig/TcR REARRANGEMENT

The V, D and J gene segments of the Ig and TcR genes are flanked by specific nucleotide sequences, which are important recognition sequences in the process of DNA rearrangement. Flanking the 3' side of the V segment, both sides of the D segment and the 5' side of the J segment are special nucleotide sequences consisting of a heptamer (seven nucleotides) separated by 12 or 23 nucleotides from a nonamer (nine nucleotides) (a). The whole sequence of heptamer + 12 or 23 bps + nonamer is known as a recognition signal sequence (RSS). These RSSs are involved in the first stage of rearrangement, the stages of which are described below.

(1) Alignment of RSS nonamers

The nonamer from an RSS with a 12-nucleotide space is brought together with the nonamer from an RSS with a 23-nucleotide space ((a) 1). The rule that an RSS with a 12-nucleotide space can bind only to an RSS with a 23-nucleotide space ensures that when the IgH and TcRβ genes rearrange, V rearranges to D and not to J.

(2) Cleavage of DNA and generation of P-nucleotides

When the nonamers are aligned, the DNA strand within the heptamers is cleaved by endonucleases and the intervening DNA is excised, or 'looped out' ((a) 2). The ends of the V and D genes now contain parts of the heptamer sequences ((a) 3).

(3) Formation and cleavage of hairpins

The two strands of DNA at the ends of the V and D genes join to form a hairpin ((b) 1). The hairpins are then cleaved at random ((b) 2) so that each DNA strand contains some nucleotides derived from the heptamer sequence ((b) 2). These nucleotides are called P-nucleotides and this process increases the variety of final DNA sequences that can be generated.

(4) Addition of N-nucleotides

The enzyme **terminal deoxynucleotide transferase** adds nucleotides at random to the ends of the DNA strands ((b) 3), further increasing variability of DNA sequences in different cells. These nucleotides are known as N-nucleotides.

(5) Filling in of DNA

Exonucleases, DNA polymerase and DNA ligase repair the DNA and complete the formation of the joint with complete nucleotide sequences on both DNA strands ((b) 4).

The example shows that generation of P- and N-nucleotides has changed the amino acid sequence of the final protein from -Arg-Val- to -Arg-Tyr-Gly-Val-Arg-Val (c). Note that the actual nucleotides and amino acids shown are for illustrative purposes only and do not represent actual sequences.

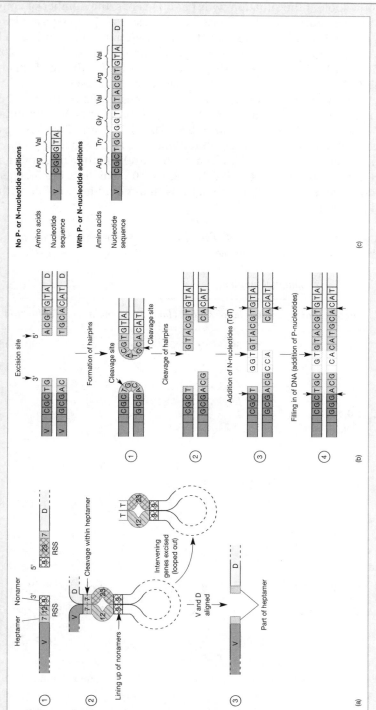

(a) Both of the DNA strands are shown as a single block. (b) The two DNA strands are shown separately. (c) Effect of P- and N-nucleotides on the amino acid sequence of the rearranged gene.

If we now look at the situation with J and D genes, the number of combinations increases dramatically. There are 51 V genes, 27 D genes and 6 J genes for the Ig H chain (Table 5.2). Each individual developing B cell selects one V_H, one D_H and one J_H gene, giving a total of $51 \times 27 \times 6 = 8262$ combinations of V, D and J. Already you can see that for the Ig H-chain, having the CDR3 made up of V + D + J segments increases the number of CDR3 variations from 51 (the number of V_H genes) to 8262 (the number of VDJ combinations). For the light chains, the number of VJ combinations are $40 \times 5 = 200$ for κ and $30 \times 4 = 120$ for λ (Table 5.2), giving a total of 320 different light chain combinations at CDR3. Since any heavy chain can associate with any light chain, the total number of combinations of H + L is now 8262 H $\times$ 320 L $= 2.6 \times 10^6$. This is a big increase on 3570, the number of H + L combinations that would be possible without D and J segments.

Table 5.2 The effect of having separate V, D and J genes on the number of different immunoglobulin and TcR variations possible

Receptor chain	Immunoglobulin							TcR				
	Heavy chain			**κ-light chain**		**λ-light chain**		**TcRα-chain**		**TcRβ-chain**		
Gene	VH	DH	JH	Vκ	Jκ	Vλ	Jλ	Vα	Jα	Vβ	Dβ	Jβ
Number of genes	51	27	6	40	5	30	4	50	70	57	2	13
Without using D or J												
Genes used	V_H			Vκ		Vλ		Vα		Vβ		
Number of chains	51			40 (= 70 light chains)		30		50		57		
Number of H + L and α + β pairs	= 3570			51 × 70				= 2850		50 × 57		
Using D and J												
Genes used	VH	DH	JH	Vκ	Jκ	Vλ	Jλ	Vα	Jα	Vβ	Dβ	Jβ
Number of chains	51 × 27 × 6 = 8262			40 × 5 = 200 (= 320 light chains)		30 × 4 = 120		50 × 70 = 3500		57 × 2 × 13 = 1482		
Number of H + L and α + β pairs				8262 × 320 = 2.6 × 10⁶					3500 × 1482 = 5.2 × 10⁶			

For the TcR, the number of combinations is increased from 2850 without the use of J and D genes to 5.2×10^6 incorporating J and D usage (see Table 5.2).

5.3.2 Additional mechanisms exist for the generation of diversity in CDR3

Although the use of J and D segments increases the number of different Ig and TcRs from thousands to millions, this still falls short of the number of different receptor specificities required to recognise 10^{11} antigen epitopes. To increase the range of specificities of Ig and TcR molecules still further, gene rearrangement involves special processes that increase the variety of sequences in the CDR3 without using extra genetic material. These processes include imprecise joining, the use of different numbers of D regions and the addition of P-nucleotides and N-nucleotides.

Imprecise joining during gene rearrangement
The mechanisms of gene rearrangement are described more fully in Box 5.2. Basically gene rearrangement involves joining the 3' end of one DNA segment to the 5' end of another (see Figure 5.5). However, this joining does not always involve the exact joining of the nucleotides that are at the end of each segment. If we take the joining of the V and D genes of the Ig heavy chain as an example, in Figure 5.6 the nucleotide at the 3' end of the V region is a guanosine and that at the 5' end of D region is an adenosine. If the joining was precise, during gene rearrangement the cytosine at the 3' end of the V region would join up with the guanosine at the 5' end of the D region. In fact, the nucleotide at the 3' end of the V region can join with nucleotides downstream of the 5' end of the D region (Figure 5.6). This imprecise joining occurs at all the rearrangements of V to D, D to J and V to J in the Ig heavy and light chain and TcR α- and β-chain genes.

An additional level of imprecision is seen in the joining of V, D and J in the Ig H and TcRα and β genes. It is not always the nucleotide at the 3' end of the V or D region that is joined to the D or J region respectively. Nucleotides upstream of the 3' end of the V and D regions can join to the D or J region (Figure 5.6).

Use of different number of D regions by the TcR β gene
A mechanism that is unique to the TcR β gene is that one T cell may use none, one or two of the D gene segments during rearrangement, offering increased diversity in the rearranged gene.

Addition of P-nucleotides
During the gene rearrangement process some nucleotides that are strictly part of an intron are randomly left attached to the V, D or J DNA segments. These are called P-nucleotides and are not removed but can contribute to the final DNA sequence coding for the protein (see Box 5.2).

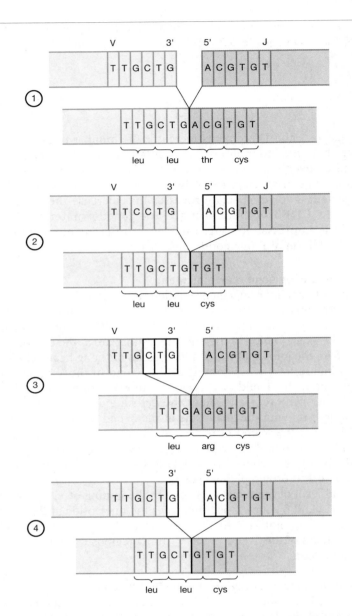

Figure 5.6 Precise and imprecise joining of DNA. During the rearrangement of V to J, V to D or D to J the joining can be precise, as shown in ①, where the 3' nucleotide of V joins to the 5' nucleotide of J. However, the joining can be imprecise: the 3' end of V may join to nucleotides downstream of the 5' end of J ②; a nucleotide upstream of the 3' end of V may join to the 5' end of J ③ or both V and J may show imprecise joining ④. Finally, nucleotides may be replaced ⑤ by special mechanisms (see Box 5.2). In this example the original nucleotides GAC were changed to ATG, changing the amino acid threonine (see ①) to a tryptophan.

Figure 5.6 Precise and imprecise joining of DNA. Continued

Addition of N-nucleotides

During the joining process some nucleotides may be lost. This means that there will be fewer amino acids in the CDR3 region of the protein. Having different numbers of amino acids in the CDR3 further increases the diversity. However, if the number of nucleotides lost is not three or a multiple of three (three nucleotides being a codon for a particular amino acid), the nucleotide sequence will potentially be out of frame and no sensible protein will be made. Two features exist to resolve this problem. Since the Ig H and TcR β genes undergo two rearrangements, V to D and D to J, it is possible, if the first rearrangement puts the nucleotide sequence out of frame, that the second rearrangement puts the sequence back in frame for reading the rest of the DNA sequence coding for the constant region. This obviously cannot happen for the Ig L and TcR α genes, where there is only one rearrangement. However, there is a mechanism that can add nucleotides to the rearranged genes and therefore replace the lost nucleotides or restore the number of nucleotides so that the nucleotide sequence is back in frame.

Developing B and T cells express an enzyme called terminal deoxynucleotide transferase (TdT) at the time they are rearranging their Ig or TcR genes. TdT has the ability to add nucleotides to a DNA strand in the absence of a DNA template (see Box 5.2). Therefore nucleotides can be added to the Ig or TcR genes during rearrangement. Although TdT has a preference for adding G and C, the nucleotides are essentially added at random. Therefore the nucleotides added are not the same ones that were lost during imprecise joining of V to D, D to J or V to J. The net effect of this process is to replace the nucleotides coded for by the germ-line DNA with nucleotides chosen at random. This is another clever way of increasing the variety of nucleotide (and hence amino acid) sequences in the CDR3 without the use of extra genetic material. N-nucleotides can be added to TcR α and β genes and the Ig H-chain genes but not the Ig L-chain genes.

The net effect of the imprecise joining mechanisms and the involvement of P- and N-nucleotides is that at every join (V to D, D to J and V to J) the nucleotides vary both in number and nature. The number of nucleotides can vary by up to 15 at each join, meaning that the number of amino acids varies by up to five. Furthermore the nucleotides, and therefore the amino acids, are randomly generated. This obviously increases the variation of CDR3 enormously (see below).

5.3.3 Somatic mutation

There is another mechanism that is used to increase the diversity of antigen-binding variations and that is somatic mutation, which is described in Section 7.4.2. This happens exclusively in B cells during their differentiation into antibody-secreting plasma cells after stimulation with antigen. During cell division, mutations arise in the Ig variable regions. These are thought to be very susceptible to mutation, although the molecular mechanisms for this are not well understood. These changes can alter the specificity or affinity of the antibody for antigen and are important for the generation of high-affinity antibody. T cells do not undergo somatic mutation.

5.3.4 Estimate of the total size of the variety of Ig and TcRs

It is difficult to estimate exactly how much the junctional diversity at CDR3 increases the number of amino acid sequences present in the CDR3 region but the potential increase is huge. If the nucleotides for just five amino acids are altered, by some being either lost or replaced, the number of different amino acid sequences that can be generated is 20^5 or about 3×10^6. In the Ig H-chain and TcR β-chain genes many more amino acids can be lost or changed. This means that when the random association of H + L or $\alpha + \beta$ chains is taken into account the number of different combinations is in excess of 10^{12}, easily enough to generate more variations of antigen receptor than the minimum estimated number of 10^{11} antigenic epitopes (Table 5.2).

Not all variations of the antigen receptors will be expressed at once: this would be impossible in a mouse that has only 10^9 T or B cells. However, the potential to generate so many different receptors for antigen means that a wide enough variety can be expressed at one time so that at least some antigens on a pathogen will be recognised by some B cells and some T cells.

5.4 Summary

- Ig and the TcR are made up of two protein chains – H and L chains for Ig, and α- and β-chains for the TcR. Any heavy chain can associate with any light chain and any α-chain can associate with any β-chain. Therefore the number of possible receptors is determined by the number of possible pairings of the two chains.

- The genes coding for the Ig and TcR chains are made up of segments. Some segments code for the C regions of the receptors and some for the V regions. The segments coding for the variable regions of the Ig heavy chains are called V_H, D_H and J_H and those coding for the Ig light chains are called V_κ and J_κ or V_λ and J_λ, depending on whether the B cell is using a κ or λ-light chain. The TcR α-chain genes coding for the variable region are called Vα and Jα and the genes coding for the β-chain are Vβ, Dβ and Jβ.

- There are many gene segments coding for V, D and J but each lymphocyte will express only one of each segment. Therefore a B cell will use one V_H, one D_H and one J_H segment for the heavy chain and one V_κ and one J_κ segment or one V_λ and one J_λ segment for the light chain. These segments are chosen at random in each B or T cell during a process called gene rearrangement. The random selection of V, D and J segments generates combinatorial diversity.

- When the Ig or TcR genes rearrange the DNA does not join precisely, so that additional variation is introduced into the V-region DNA sequence at the site of the junction of V to D or J and D to J. This additional variation is called junctional diversity and greatly increases the number of different antigen receptor specificities.

- B cells, but not T cells, somatically mutate their Ig V-region genes during affinity maturation, adding further diversity to the number of Ig specificities.

- The different mechanisms used to generate diversity ensure that enough different receptors are generated to interact with any antigen.

CHAPTER SIX

Anatomy of the immune system

Learning objectives

To understand the problems of generating immune responses *in vivo*. To learn the anatomy of the lymphoid system and how this promotes the generation of immune responses. To know how cells of the immune system move throughout the body and within tissues and the molecular basis controlling this movement.

Key topics

- Requirements of the immune system *in vivo* and the need to generate specific immune responses
- Anatomy of the lymphoid system
  - The lymphatic system
  - Lymph nodes
  - Spleen
  - Mucosal associated lymphoid tissue (MALT)
- Lymphocyte recirculation

6.1 Requirements of the immune system *in vivo*

The job description of the immune system *in vivo* is very simple: to protect every tissue and organ in the body against any pathogen. How to achieve this is not quite so simple. The immune system does not know what pathogens are waiting to infect an individual or where a particular pathogen will enter the body. It therefore has to be ready to respond to any situation.

The ultimate aim of an immune response is to eliminate or neutralise threats to the body posed by an infectious agent (pathogen). Sometimes this can be achieved by the innate immune system; for instance, phagocytes may engulf and destroy infectious bacteria. However, in many situations the innate immune system alone cannot cope and requires help from the specific immune system. The specific immune system consists essentially of the B lymphocytes and CD4 and CD8 T lymphocytes described in Chapters 3 and 4 and therefore a specific immune response involves one or more of these cell types.

6.2 The types of specific immune response

There are basically three ways in which a specific immune response can contribute to the elimination of pathogens:

- **The generation of antibody.** Antibody can interact with a large variety of components of the innate immune system to help neutralise or eliminate a pathogen. Antibody is produced by **plasma cells**, which are themselves derived from B lymphocytes. Therefore antibody production requires the differentiation of antigen-specific B lymphocytes into plasma cells that produce antibody against the antigen.

- **The production of CD8 cytotoxic T cells (Tcs).** Tcs can kill other cells that are expressing antigen on their class I MHC molecules. This response is particularly important in killing virally infected cells before the virus has a chance to replicate inside the cells.

- **Delayed type hypersensitivity (DTH) responses.** Most tissues contain some macrophages. These macrophages can respond to pathogens or their products and be activated to kill many infectious agents. Some pathogens, however, have evolved mechanisms to resist killing by these tissue macrophages; in many cases the pathogenic organism is actually able to survive and replicate within the tissue macrophages themselves. Additionally there may simply not be enough macrophages to deal with the number of infectious particles. This is where a DTH response is required. A DTH response involves two important factors: monocytes are recruited from the blood to the site of infection and both the recruited monocytes and the resident tissue macrophages are activated so that they are better able to kill the pathogen (see Section 8.5.3).

All of these specific immune responses are dependent on CD4 T cells (Figure 6.1). Without CD4 T cells antibody cannot be generated against most antigens, Tcs cannot be produced against many viruses and DTH reactions do not occur. This explains to a large extent the devastating effect of the AIDS virus on the immune system. One of the main effects of the AIDS virus is to cause the loss of CD4 T cells, resulting in the reduced ability to generate any type of acquired response and thus profound immunosuppression.

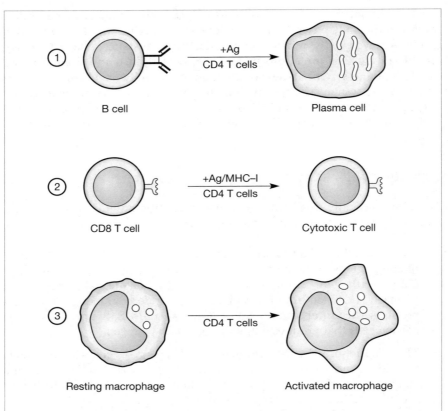

Figure 6.1 CD4 Th are required for all types of specific immune responses. Th are required for B cells to differentiate into plasma cells ①, for naive CD8 T cells to become effector Tcs ② and for the activation of macrophages in a DTH response ③.

Therefore, during a specific immune response, CD4 T cells must interact with many other cell types including B cells, CD8 T cells and macrophages; the specific interactions depend on the nature of the immune response. Another important consideration is that if this is the first infection with a particular pathogen, B cells and T cells specific for the antigen will be quite rare. This poses two problems for specific immune responses. The first of these is that specific immune responses must be acquired.

6.2.1 Specific immunity must be acquired

The reason why specific immune responses must be acquired is best explained if we look at the situation with antibody. It was estimated in Chapter 5 that there will be in excess of 10^{11} different antigenic epitopes and in excess of 10^{11} different antibodies will be needed to recognise all these different epitopes. At first sight it might seem easiest just to make all of these different antibodies and then, when an infection occurs, antibodies with specificity for the antigens on the pathogen will bind to those antigens.

This is not possible for the following reason. It has been calculated that a protective concentration of antibody, in terms of eliminating or neutralising an infectious agent, is 10 ng/ml. Therefore, to maintain this level of antibody for 10^{11} different antibody specificities in the blood of an adult human whose blood volume is 5 L would require:

5000   volume of blood (mls) $\times$
10^{11} number of different antibody molecules required $\times$
10   protective level of antibody (ng/ml)
$= 5 \times 10^{15}$ ng $= 5 \times 10^6$ g or 5000 kg of antibody

This is approximately 5 tons of antibody! It is clearly not possible to have this amount of antibody present all the time. Another alternative strategy is therefore required to provide antibody against a particular pathogen when that antibody is required. Fortunately we are exposed to only a limited number of infections at one time and therefore only need to make antibody against a minute fraction of the total number of antigenic epitopes that exist. However, this poses a problem. Since the immune system does not know which antigens it will encounter, how does it select which antibodies to make? The only feasible way to do this is to make antibody specific for an antigen **in response to the presence of the antigen**. In other words the antibody response is **acquired** after the immune system is exposed to antigen.

Just as it is not possible to have pre-existing antibody against all possible antigens, it is not possible to have enough antibody-producing cells prior to exposure to antigen. The antibody-producing cells have to be produced after exposure to antigen. The situation is the same for T lymphocytes; although there will be some T cells with specificity for antigens derived from an infectious agent there will not be enough to deal with the infection and it will be necessary to generate new T cells specific for the antigen after exposure to antigen. Therefore specific immunity, whether it is antibody production or T lymphocyte responses, is acquired after exposure to antigen. The need to acquire specific immunity is another feature that distinguishes specific immunity from innate immunity. Innate immunity exists in the absence of infectious organisms, although the level of certain components can increase in the presence of an infectious agent.

6.2.2 Cellular interactions are required during specific immune responses

The requirement to generate specific immune responses after exposure to antigen poses another major challenge for the immune system *in vivo*. Simply put, how do you arrange things so that extremely rare antigen-specific B cells and T cells can interact both with antigen and with each other? If you consider a human being, it is quite a big place for small cells such as lymphocytes, and they need to be in the same place at the same time to encounter antigen. Normally we are infected with relatively small numbers of infectious particles and so the amount of antigen is quite low. Also, as described in Chapter 1, infectious agents, and the antigens associated with them, can arrive by various routes. Finally, antigen-presenting cells are required and essential for presenting antigen on MHC to T cells. So how does the immune system bring together rare antigen-specific T and B cells, antigen-presenting cells and antigen to a single anatomical site so that antigen stimulation and the cellular interactions involved in generating effector cells can occur?

Basically there are two strategies to achieve this. The first is to have collections of lymphocytes and the other cell types required to generate an immune response located in the sites of the body where infection is most likely to occur – these are the mucosa of the GI, respiratory and GU tracts. These collections of lymphoid cells are called the mucosal associated lymphoid tissue, or MALT, and are described in Section 6.3.4. The second strategy is to have lymphocytes circulate throughout the body and have specialised anatomical structures designed to capture antigen and enable lymphocytes specific for the antigen to meet the antigen and get together with the other cell types required to generate an immune response. These structures are the spleen and lymph nodes (Figure 6.2). By

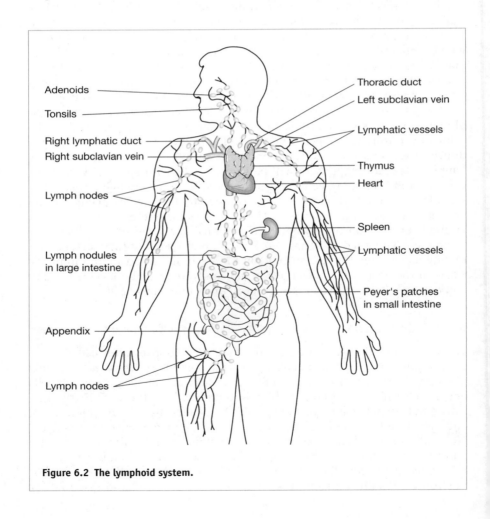

Figure 6.2 The lymphoid system.

constantly recirculating through the spleen and lymph nodes, antigen-specific lymphocytes patrol the whole body. If antigen is present in a specific lymph node or the spleen, cells can stop recirculating and antigen-specific cells accumulate at the site of antigen.

6.3 Anatomy of the lymphoid system

The immune system consists of a series of organs and the vessels that connect them (Figure 6.2). Organs of the immune system have been divided into **primary** and **secondary** lymphoid tissue. The two primary lymphoid organs in most mammals are the thymus and bone marrow. The term 'primary' is used because these are organs in which naive lymphocytes develop from bone-marrow precursors; B cells develop in the bone marrow and T cells in the thymus. This is an antigen-independent process and is described in Chapter 11. Once lymphocytes are formed they migrate to the secondary lymphoid organs and tissue. The secondary lymphoid tissue consists of the lymph nodes, spleen and MALT.

Other specialised structures of the lymphoid system are the **lymphatic vessels**. These are somewhat similar to blood vessels and connect most tissues in the body with lymph nodes and eventually with the bloodstream. Lymphocytes recirculate constantly through blood and lymphatic vessels. It should be noted that the spleen is an important lymphoid organ that does not have connections with lymphatic vessels; as will be seen later, it is primarily devoted to dealing with antigen in the bloodstream.

6.3.1 The lymphatic system

The ends of lymphatic vessels are found in most tissues and internal organs (Figure 6.3). They are somewhat similar to blood vessels but the content of fluid and cells differs between blood and lymph. Lymph is derived from the interstitial fluid surrounding cells in any tissue or organ and is lower in protein content than blood plasma. Certain cells also enter the lymphatic vessels from tissue; these are lymphocytes and tissue dendritic cells but lymph does not normally contain red blood cells, monocytes or macrophages, or neutrophils, eosinophils or basophils. The smallest lymphatic vessels are called lymphatic capillaries and these eventually join and become larger lymphatic vessels, which eventually re-enter the blood stream via the largest lymphatic vessel, the thoracic duct (Figure 6.2).

6.3.2 Lymph nodes

At regular intervals in the lymphatic system a number of lymphatic vessels meet at specialised structures called **lymph nodes** (Figure 6.4 and Plate 9). Lymph nodes are kidney-shaped structures; in humans they vary in size from a few millimetres to two cm in length. They are surrounded by a capsule consisting of connective tissue and are grey-white in colour. Immediately under the capsule is an area called the sub-scapular sinus. The lymphatic vessels that bring lymph fluid into the lymph nodes are called **afferent lymphatic vessels**

Figure 6.3 The lymphatic system. Lymphatic capillaries drain most tissues in the body. They join afferent lymphatic vessels, which eventually drain into lymph nodes. Lymphatic vessels leaving lymph nodes are called efferent lymphatic vessels, which may drain into other lymph nodes. Eventually efferent lymphatic vessels join to become larger and drain into the bloodstream via two major lymphatic ducts, the thoracic duct and the right lymphatic duct.

and they pierce the capsule and empty their contents into the sub-capsular sinus. The lymph fluid filters through the lymph node, carrying with it lymphocytes and antigen. The lymph fluid eventually leaves the lymph node via the efferent lymphatic vessel, which exits the lymph node at a region called the hilum. The hilum is also the site of the lymph node where blood vessels enter and leave the lymph node.

Figure 6.4 Lymph node. The innermost part of the lymph node is called the medulla and is where plasma cells secreting antibody are located. The outer part of the lymph node consists of the cortex, which is primarily a B-cell area containing primary follicles and secondary follicles with germinal centres. The paracortex is located between the cortex and medulla and contains mainly T cells.

Morphologically lymph nodes consist of a reticular framework made up of fibres and epithelial cells which provide structural support for the organ. Like many glands lymph nodes are divided into an outer cortex and an inner medulla. Within the cortex can be seen circular aggregates of lymphocytes which, under haematoxylin and eosin (H & E) staining, may or may not have a lighter staining centre. The aggregates with a darker staining area are called primary follicles and consist mostly of unstimulated B cells. The aggregates with a lighter staining centre are called germinal centres or secondary follicles and develop following antigen stimulation (see Section 7.4). Germinal centres are where plasma cells and memory B cells are generated. Between the follicles is an area called the paracortex, which contains densely packed T cells together with some dendritic cells and B cells. The medulla is less densely packed with cells and consists mainly of strands, or cords, of cells containing T cells, macrophages and large numbers of plasma cells.

6.3.3 The spleen

The spleen is located on the left side of the abdomen and in humans is oval with a length of about 12 cm and a diameter of 5 cm. It is surrounded by a connective tissue capsule. Examination of a cut spleen under the naked eye reveals a predominantly red surface with greyish white specks. These two areas are called the red pulp and the white pulp. The red and white pulp

Figure 6.5 Spleen. (a) Transverse section of spleen. Arrows indicate direction of blood flow. Most of the spleen consists of red pulp, which is primarily involved in removing old erythrocytes and platelets. The lymphoid area of the spleen is the white pulp, which surrounds the central arterioles. (b) Expanded view of white pulp. The area immediately surrounding the central arteriole is the peri-arteriolar lymphoid sheath (PALS), which contains mostly T cells. Adjacent to the PALS is the B-cell area containing primary follicles and secondary follicle with germinal centres. Between the white pulp and the red pulp is the area called the marginal zone, which contains B cells and macrophages.

areas of the spleen can be clearly seen histologically (Figure 6.5 and Plate 9). The spleen is surrounded by a collagenous capsule and bundles of fibres called trabeculae extend from the capsule into the internal part of the spleen, partially separating it into compartments. The structure of the spleen is further maintained by a reticular framework. The function of the red pulp is to filter the blood and remove aged or damaged red blood cells and other debris; this function is described in Box 6.1.

BOX 6.1: THE SPLEEN AND RED BLOOD CELLS

The main function of the red pulp of the spleen is to filter the blood and remove old (effete) and damaged red blood cells. Not surprisingly the spleen has an extensive blood supply. Blood enters the spleen via the splenic artery, which divides into many arteries called central arteries. These arteries become thinner arterioles, which eventually enter the red pulp. The red pulp contains thin-walled blood vessels called venous sinusoids and in between these sinusoids are areas called splenic cords. Blood cells are emptied out of the arterioles directly into the splenic cords. To re-enter the blood circulation, blood cells must traverse the splenic cords and enter the venous sinusoids. Although the walls of the venous sinusoids are incomplete they are lined with macrophages and the splenic cords are full of macrophages. These macrophages recognise and phagocytose old or damaged red cells and platelets, preventing their re-entry into the blood. Blood cells must squeeze through the walls of the discontinuous endothelium of the venous sinusoids to re-enter the bloodstream. Older red cells, whose membranes are less elastic, are unable to do so, thereby providing an additional mechanism by which the spleen removes aged red blood cells from the circulation. The venous sinusoids eventually join to form larger veins, which eventually leave the spleen via the splenic vein.

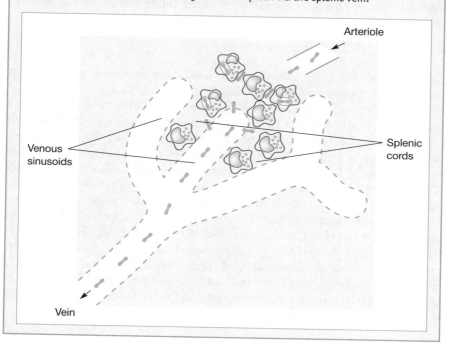

The immune function of the spleen is performed primarily in the white pulp and the spleen is designed to filter antigen from the blood so that T and B cells can encounter the antigen and respond to it. Blood enters the spleen via the splenic artery, which divides into trabecular arteries; these in turn divide into many smaller arteries called the central arteries. The central arteries become thinner arterioles, which eventually enter the red pulp. Surrounding the central arteries are sheaths of lymphoid cells consisting of T cells, B cells, macrophages and dendritic cells. T cells are usually located in the area immediately surrounding the arteries, which is called the peri-arteriolar lymphoid sheath, or PALS. The area immediately surrounding the PALS is called the marginal zone and contains B cells and macrophages. Structures called primary follicles and germinal centres are found at regular intervals along the sheaths; the germinal centres are the main sites of antibody production. As in the lymph node, the primary follicles represent collections of unstimulated recirculating naive B cells and the germinal centres form after antigen stimulation.

6.3.4 MALT

MALT consists of more or less structured lymphoid tissue distributed throughout the mucosa of the GI, respiratory and GU tracts. The more structured elements of MALT include the tonsils, adenoids, appendix and Peyer's patches, which line the ileum (Figure 6.2). Less structured lymphoid follicles are also found in the lamina propria of the intestine and the mucosa of the respiratory and GU tracts. MALT contains lymphoid follicles like those seen in the spleen and lymph nodes (Plate 10).

6.4 Lymphocyte recirculation

As mentioned above, the second way in which rare antigen-specific cells can meet with each other and interact with antigen and APCs is through lymphocyte recirculation. Lymphocytes differ from other leucocytes in that they constantly migrate between blood, the lymphatic vessels and the lymphoid organs. This process is called **lymphocyte recirculation** and the extent of this cell movement is quite remarkable. Lymphocytes circulate in the bloodstream for an average of 30 minutes. Approximately 45% of the cells enter the spleen where they have a transit time of about 5 hours before returning to the blood stream via the splenic vein. However, lymphocytes are unique in having the ability to migrate directly from the bloodstream into lymph nodes. They do this by leaving the blood in **high endothelial venules** (HEVs). High endothelial venules are specialised blood vessels that have a cuboidal (high) endothelium (Plate 9). About 40% of the lymphocytes in the bloodstream enter lymph nodes via this route and reside in the lymph nodes for 12 hours or so before leaving via the efferent lymphatics, which eventually transport the lymphocytes to the thoracic duct and back into the bloodstream. Lymphocytes also enter lymphatic vessels in tissues and can travel along the lymphatic vessels to lymph nodes. About 15% of lymphocytes entering a lymph node enter via the lymphatic vessels, and 85% enter across HEVs from the blood. Because of this extensive recirculation, lymphocytes migrate from blood to spleen and lymph nodes one or two times/day. The final few per cent of lymphocytes enter the MALT or non-lymphoid tissues.

The migration of lymphocytes from blood into lymph nodes involves the same basic processes as seen with leucocytes entering sites of inflammation (see Chapter 2). The blood flow in the lymph node slows because of vasodilation. However, this vasodilation is not caused by inflammation but vasodilation occurs because the high endothelial venules that are the site of migration from blood to lymph nodes are located where small-diameter capillaries become larger-diameter venules. This change in diameter slows the blood in the same way as the change in diameter caused by inflammation. This slowing of blood flow allows the lymphocytes to roll along the vessel wall. The subsequent stages of weak binding of adhesion molecules on the lymphocytes to those on endothelial cells, activation of adhesion molecules, firm attachment and crossing of the endothelium are the same processes as seen for leucocytes entering sites of inflammation, although the adhesion molecules and chemokines involved are different (see Table 6.1). The HEVs are located in the paracortex of the lymph node so that the lymphocytes enter the lymph node in a T-cell area. Since B cells also leave the bloodstream and enter the lymph node they must then migrate to the B-cell area, the cortex and follicles. The movement of the B-cells to the follicles is promoted by the chemokine B lymphocyte chemo-attractant (BLC), which is produced by stromal cells of the B-cell areas and follicular dendritic cells and binds to the chemokine receptor CXCR5, which is expressed on the B cells.

Table 6.1 Adhesion molecules involved in migration of lymphocytes to lymphoid tissue

Adhesion molecule	Lymphocyte distribution	Endothelial cell ligand	Type of migration affected
L-selectin	B + T	GlyCAM-1	Migration to peripheral lymph node
		MadCAM-1	Migration to mucosal lymphoid tissue
$\alpha_4\beta_7$ integrin	T	MadCAM-1	Migration across mucosal HEV
$\alpha_L\beta_2$ integrin	T	ICAM-1,2,3	Migration across HEV

6.5 Summary

- There are three types of specific immune response: the production of antibody, the generation of CD8 cytotoxic T cells, and delayed hyper-sensitivity responses.

- Immune responses must be generated in response to infection and this requires the interaction of rare antigen-specific cells with each other and with antigen.

- The immune system has specialised anatomical structures that promote the association of antigen with antigen-presenting cells and antigen-specific B and T cells.

- These structures include the lymphatic vessels, which connect together lymph nodes and eventually join with the blood system.

- The major lymphoid organs are the lymph nodes and spleen, together with a vast collection of lymphoid tissue associated with the mucosa, which is called mucosal associated lymphoid tissue or MALT.

- MALT consists of aggregates of structured lymphoid tissue such as the tonsils, appendix and Peyer's patches and loose aggregates of lymphoid cells called nodules.

- Lymphocytes recirculate extensively between the bloodstream and lymph nodes and spleen. This recirculation is controlled by adhesion molecules and chemokines and increases the chances of lymphocytes meeting their specific antigen.

CHAPTER SEVEN

Anatomical and cellular aspects of antibody production

Learning objectives

To learn how antibody is produced *in vivo* in response to antigen. To know about the role of CD4 T cells and cytokines in antibody production. To understand how B cells switch the class of antibody they make and increase the affinity of their antibody during a response to antigen and why this makes the response much more efficient. To know that B cells can become either plasma cells or memory B cells.

Key topics

- Requirements for antibody production
- CD4 T cells and cytokines in antibody production
- B lymphocytes and antibody production
  - ○ Clonal selection
  - ○ Class switch
  - ○ Affinity maturation
  - ○ Differentiation into plasma or memory cells
- The advantages of class switch and affinity maturation

7.1 Overview of antibody production

As mentioned in Chapter 6, antibody is produced by plasma cells that differentiate from antigen-specific B cells. B lymphocytes that have never encountered antigen before have IgM and IgD on their surface (see Figure 3.11). They are able to recognise and bind antigen through IgM and IgD on their surface but at this stage cannot secrete antibody. For an individual B cell the IgM and IgD have identical variable Ig Vh regions and identical light chains and therefore the same antigen specificity. However, different B cells will have specificity for different antigens. To become a plasma cell, a B cell must undergo a process of **cell differentiation**. Many cell types, not just those of the immune system, undergo differentiation, usually during development or as a result of external signals such as hormones. Cell differentiation at its most basic involves changes in gene expression that alter the function of a cell. B cells must undergo a number of changes in their differentiation into plasma cells. This includes changing from a cell with Ig on its surface to a cell that can secrete enormous amounts of antibody. Plasma cells are basically cellular antibody-producing factories. B lymphocytes also usually go through two specialised differentiation processes during their differentiation into plasma cells. These processes are designed to increase the quality and range of antibodies produced and are:

- **Affinity maturation of antibody.** This is a specialised process that results in antibody with a high affinity for antigen and that is more effective at dealing with pathogens.

- **Antibody class switch.** As described in Chapter 3, there are many different classes of antibody, IgG, IgA and IgE, in addition to IgM and IgD. These classes of antibody have different biological functions. During differentiation into plasma cells, B cells can change the antibody on their cell surface from IgM and IgD to one of the other classes of antibody, which means that when they differentiate into plasma cells they secrete different classes of antibody with different biological function. It should be noted that an individual plasma cell will secrete only one class of antibody.

It can be seen that the differentiation of B lymphocytes into plasma cells is a complex process that must be carefully regulated. An important cell in the regulation of B-cell differentiation is the CD4 T cell. The CD4 T cells that are involved in antibody production are called helper T cells, or Th for short. They were named helper T cells because although they didn't make antibody themselves, they were required for B cells to make antibody, i.e. helped in antibody production. Most antibody production is dependent on Th although some antigens can stimulate antibody production without the need for Th (Box 7.1). However, just as B cells that have never seen antigen before cannot secrete antibody and have to differentiate into plasma cells to do so, CD4 T cells that have never seen antigen before have no helper function and must undergo their process of differentiation into Th following stimulation by antigen peptide/class II MHC.

Another important feature of specific immune responses is the need for antigen-specific cells to proliferate after appropriate stimulation by antigen. If you are infected with a microbe for the first time you will have very few lymphocytes that are specific for the microbial antigens. Therefore it is an important requirement to increase the number of antigen-specific cells as quickly as possible and this is achieved by rapid cell proliferation. Because

B cells, like T cells, require two stimuli for activation by most antigens. The first stimulus is recognition of antigen by membrane Ig (mIg) and the second stimulus comes from the CD4 T cell. There are certain antigens, however, that can deliver a strong enough signal through the mIg complex to activate B cells without any help from CD4 T cells. These are called T-independent antigens because they can stimulate antibody production without T-cell help. These antigens are usually polysaccharides with highly repetitive epitopes causing extensive cross-linking of the mIg and strong signalling (see figure). However, responses to these antigens in the absence of T cells involve only IgM production and there is no class switching or affinity maturation. Most antigens cannot cause such strong signalling in B cells and the cells require additional signals or co-stimuli and T-cell help. The role of Th is to provide additional signals for the initial activation of B cells and to control the subsequent proliferation and differentiation of B cells both by secreting cytokines and by delivering signals to B cells by cell-to-cell contact between the B cell and CD4 T cell.

only lymphocytes that are specific for antigen on microbes are stimulated to proliferate, this process has been called **clonal selection** and the proliferation of antigen-specific cells is called **clonal proliferation**. The principle of clonal selection and expansion is shown in Figure 7.1 for a B cell. T cells will undergo similar clonal selection and expansion after stimulation by antigen/MHC.

Therefore the production of antibody can be divided into a number of stages:

1. Presentation of antigen to, and stimulation of, antigen-specific CD4 T cells to proliferate and differentiate into Th.

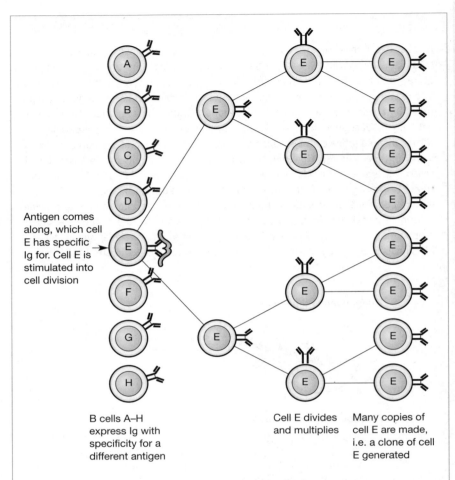

Antigen comes along, which cell E has specific Ig for. Cell E is stimulated into cell division

B cells A–H express Ig with specificity for a different antigen

Cell E divides and multiplies

Many copies of cell E are made, i.e. a clone of cell E generated

Figure 7.1 Clonal selection and expansion in B cells. The B cells labelled A–H all have Ig on their surface but the Ig on each B cell is specific for a different antigenic epitope. When antigen comes along, only B cell E has Ig that can bind to an epitope on the antigen. Binding of the antigen to the Ig on B cell E stimulates the B cell to divide many times, forming a clone of B cells derived from E.

2. Stimulation of B cells by antigen and interaction with Th.
3. Proliferation of B cells and their differentiation into plasma or memory cells.

These events occur primarily in lymph nodes, spleen and mucosal lymphoid tissue. The fundamental ways in which the spleen and lymph nodes operate to achieve these events are quite similar although some of the details differ. A major difference between lymph nodes and spleen is that antigen is brought to the lymph nodes by lymphatic vessels, whereas antigen in the spleen comes from the blood. Therefore antigen located in tissues enters the lymphatics and stimulates responses in the lymph nodes draining the tissue. Antigen in the blood enters the spleen and stimulates immune responses there.

The first, and critical, stage in any adaptive immune response is the stimulation of antigen-specific CD4 T cells to become Th. This occurs in lymph nodes and the spleen and in organised mucosal tissue, the latter of which will be described in Section 7.5. When we are infected with a microbe this occurs in various tissue site, e.g. under the skin or sometimes in the blood. Therefore antigen has to get from the site of infection to a lymph node or to the spleen to be able to activate CD4 T cells to proliferate and differentiate into Ths. The production of Th can therefore be considered in three stages:

1. Delivery of antigen to lymph node or spleen.

2. Activation of antigen-specific CD4 T cells.

3. Proliferation of activated CD4 T cells and differentiation into Th.

7.2.1 Delivery of antigen to lymph nodes or spleen

When a tissue is infected by a microbe, the microbe will initially encounter cells of the innate immune system, such as tissue macrophages described in Chapter 2; this may initiate an inflammatory response. Another cell type found in tissues plays a vital role in stimulating CD4 T cells and initiating an immune response. This cell is called a **dendritic cell (DC)**. DCs are bone-marrow-derived cells that are found in almost all tissues. DCs in non-lymphoid tissues are called tissue DCs; Langerhans cells are a specialised type of tissue DC found in the skin (Plate 11). Tissue DCs are highly efficient at taking up antigen by pinocytosis, receptor-mediated endocytosis (using the mannose receptor) or phagocytosis. Tissue DCs can also recognise microbial products through the expression of toll-like receptors (TLRs – see Chapter 2) and become activated as a consequence of this. When they are activated, DCs are stimulated to leave the tissue site and enter the draining lymphatic vessels in which they will be carried to the local lymph node. The migration of dendritic cells to the lymphatic vessels is induced by a chemokine called the Epstein–Barr virus-induced receptor ligand chemokine (ELC), which binds the chemokine receptor CCR7, which is upregulated on the activated DCs. As they migrate from the tissue to the lymph node the DCs process the antigen and express antigen fragments on their class II MHC for recognition by CD T cells. Once in the lymph node the dendritic cells migrate to the T-cell area, the paracortex, where they can encounter CD4 T cells. Antigen derived from microbes at the site of infection can also enter the lymphatic ducts in free form and be carried to the local lymph nodes. As the antigen percolates through the lymph node it is picked up by DCs in the paracortex, where it can be processed and antigenic fragments presented to T cells on class II MHC.

Antigen can also arrive in the spleen in two ways. DCs travel from tissue via the blood to the spleen carrying antigen for presentation to CD4 T cells. Alternatively many of the central arteries in the spleen terminate near, or in, the marginal zone. Blood containing antigen is emptied into this area and some of the antigen enters the PALS area where it can be picked up and processed by dendritic cells for presentation to CD4 T cells (Figure 7.2).

7.2.2 Activation of CD4 T cells (0–5 days after antigen)

The activation of CD4 T cells is one of the most crucial early events in initiating a specific immune response. The ideal situation is that the CD4 T cell should be able to recognise and respond to as few molecules of antigen/class II MHC as possible so that the response can be initiated soon after infection. However, it is important that the response is controlled so that only foreign antigens are responded to. One of the mechanisms by which the activation of CD4 T cells is controlled is to have a requirement that the CD4 T cell must receive two signals to be activated. The first signal is generated by the TcR on the CD4 T cell surface recognising antigen in association with class II MHC. The cell bearing the antigen on its class II MHC is called an **antigen-presenting cell**, or APC. The requirement for recognition of antigen means that CD4 T cells are activated only in the presence of antigen to which they are specific; there would be no point in activating the cells if their antigen was not present.

The second requirement for the activation of CD4 T cells is that other molecules on the T cell must bind to molecules on the APC. When these other molecules on the T cell are bound they deliver another signal to the T cell and this other signal is known as a **co-stimulus** (see Figure 7.3). The most important co-stimulatory molecule on the T cell is called CD28. This can bind to either of two molecules called CD80 and CD86 (these were formerly called B7.1 and B7.2 respectively). If the TcR on a CD4 T cell binds antigen/class II MHC and CD28 binds to CD80 or CD86, the CD4 T cell will receive its two signals and be activated. In the absence of a co-stimulatory signal the CD4 T cell will not be activated and no response will occur. However, mice lacking CD28 can still generate some Th although not as effectively as normal mice, indicating that other molecules can act as co-stimuli.

Therefore an APC must express both class II MHC and CD80 or CD86 to be able to optimally activate a resting CD4 T cell. The expression of CD80 and CD86 itself is carefully controlled so that CD4 T cells only receive the co-stimulus through CD28 and are activated when it is appropriate. The tissue dendritic cells described above express some class II MHC and co-stimulatory molecules but are not very good at activating CD4 T cells. However, the DCs that have migrated to lymph nodes or the spleen following activation by microbial products have upregulated CD80 and CD86 in addition to upregulating class II MHC and have now become cells that are very good at activating CD4 T cells.

Figure 7.2 Transport of antigen to the spleen and lymph node. In this example, bacteria have infected a tissue site. Antigens shed from the bacteria can enter the draining lymphatics ① to be transported to the draining lymph node where they can directly stimulate B cells ② or be taken up by interdigitating dendritic cells within the paracortex of the lymph node, processed and antigenic peptides presented to CD4 T cells by class II MHC on the dendritic cells ③. Alternatively antigen in tissues can be taken up by tissue dendritic cells ④ that are stimulated to enter the lymphatics and migrate to the draining lymph node where they can present antigenic peptides to CD4 T cells ⑤. If there is damage to blood vessels, antigens enter the blood ⑥ and are transported to the spleen. Here they enter the white pulp ⑦ and can stimulate B cells or be taken up by interdigitating dendritic cells and antigenic peptides presented to CD4 T cells.

Figure 7.3 CD4 T cells require two signals to be activated. Signal 1 is provided by the TcR recognising Ag/class II MHC. The second signal, known as a co-stimulus, is provided by the cell presenting the antigen. The most important co-stimulus is provided by CD80 or CD86 on the APC, which binds to CD28 on the CD4 T cell.

Figure 7.4 Formation of the immunological synapse. (a) Initially the TcRs on the CD4 T cell that come into contact with class II MHC/peptide on the APCs are randomly distributed on the respective cell surfaces. CD4 also binds to class II MHC. ICAM-1 on the APC binds to LFA-1 on the T cell promoting adhesion between the two cell types. (b) Following signalling from the TcR/CD3 complex, cytoskeleton event results in reorganisation of molecules on the cell membrane to form supra-molecular activation clusters (SMACs) with TcRs and associated signalling molecules located in the central region (c-SMAC) and the LFA-1/ICAM-1 located in the periphery (p-SMAC) of the synapse. Other molecules such as CD28 also locate to the c-SMAC to provide additional signalling.

When the DCs have migrated to the T cell areas of the lymph node they secrete the chemokine ELC. ELC can bind to its receptor CCR7, which is expressed by T cells, and therefore the T cells are attracted to the dendritic cells and contact between the two cell types is promoted. The attraction of T cells to dendritic cells is not antigen-specific. A particular dendritic cell will express many different antigenic peptides on all its class II MHC molecules. However, most of the CD4 T cells that come into contact with the dendritic cells will not have TCRs that are specific for any of the antigen/class II MHC expressed by the dendritic cell; there will only be rare antigen-specific CD4 T cells that come into contact with the dendritic cell. The challenge is to maximise the activation of these antigen-specific CD4 T cells by small amounts of antigen without inadvertently activating CD4 T cells with the wrong specificity. This is achieved in part through co-stimulation described above and in part through the formation of a specialised intercellular communication structure between the dendritic cell and the CD4 T cell known as the **immunological synapse** (Figure 7.4). The term 'immunological synapse' was coined because the interaction between the T cell and dendritic cell was likened to that between neurons, which occurs via neuronal synapses.

Immunological synapse

CD4 T cells have on their surface the integrin LFA-1 whose ligand is ICAM-1. When a CD4 T cell comes into contact with a dendritic cell the CD4 T cell will 'scan' the dendritic cell to see whether it is expressing antigen for which the TCR on the T cell is specific. If the TCR does not recognise antigen it will dissociate from the APC and continue travelling through the lymph node or spleen and repeat the scanning process with other DCs. If, however, the TCR recognises antigen being presented by the DC the TCR will bind to the antigen/class II MHC. It is also important for TCR-mediated signalling that CD4 molecules on the T cell bind to class II/MHC. As a consequence of TCR recognition of antigen, signalling events occur in the T cell that result in activation of LFA-1 so it now binds ICAM-1 on the DC with high affinity. This stabilises the binding of the T cell to the dendritic cell and allows more TCR molecules on the T cell to bind to antigen/class II MHC on the dendritic cell. There is also recruitment of cytoskeletal elements within the T cell that results in redistribution of the molecules on the T cell surface into what are known as supramolecular activation clusters (SMACs). This results in SMACs where the TCRs and their associated CD3 complex (see Section 4.2), CD4 and CD28 form a central cluster surrounded by a peripheral ring of LFA-1. This clustering of the TCR/CD3 complex, CD4 and CD28 maximises the delivery of both signal 1, through the TCR/CD3 complex, and signal 2, through CD28, to the CD4 T cell, resulting in its activation.

7.2.3 Proliferation and differentiation of CD4 T cells

After activation by antigen-presenting DCs the CD4 T cells are stimulated to proliferate. The proliferation of CD4 T cells is driven by cytokines. Activated CD4 T cells produce a cytokine called interleukin-2 (IL-2). IL-2 is a protein of molecular mass 15 kDa and is primarily a growth factor. It can stimulate the proliferation of CD4 T cells, CD8 T cells and B cells. Upon stimulation through the TcR, CD4 T cells are induced to express receptors for IL-2 (IL-2R). Since the same CD4 T cell can both make IL-2 and express the IL-2R, IL-2 is able to act in an autocrine manner. IL-2 can also act in a paracrine manner and bind to IL-2Rs on neighbouring cells that have been stimulated through antigen recognition to express IL-2Rs (Figure 7.5). Through the autocrine and paracrine actions of IL-2, CD-4 T cells that have recognised antigen undergo extensive proliferation so that the number of antigen-specific CD4 T cells is increased dramatically. This proliferation peaks 3–4 days after initial contact with the antigen and, although it is difficult to measure exact figures *in vivo*, eventually results in a 10,000- to 100,000-fold increase in the number of antigen-specific CD4 T cells.

Following the period of proliferation, the CD4 T cells begin to differentiate and acquire the functions they need to act as Th cells. The main function that Th cells acquire is the ability to secrete a variety of different cytokines (Figure 7.5). There are also changes in the molecules expressed on the surface of the T cells and, through the expression of these molecules and the secretion of cytokines, the Th cells are able to control the production of different effector responses. The Th migrate to the boundary of the paracortex and the cortex, where they can interact with B cells.

7.3 Stimulation of B cells by antigen and interaction with Th (2–4 days after antigen)

Antigen filtering through the lymph node cortex percolates through the follicles, where it can be captured by macrophages (Figure 7.6a). If the antigen on the macrophage comes into contact with B cells whose Ig is specific for the antigen, two events occur. One is that the B cells take up the antigen, process it and present antigenic peptides on their class II MHC. The B cells are also partially activated by antigen and migrate to the junction of the cortex and paracortex, where they encounter Th cells that have migrated to the edge of the paracortex. The interaction between the B and Th cells involves both direct cell–cell contact and the production of cytokines (Figure 7.7). The activated B cells express increased class II MHC on their cell surface and so are able to present antigen to the Th. The Th, that are specific for the antigen/class II MHC presented by the B cells recognise the antigen/class II MHC through their TcR and are stimulated to produce cytokines. This ensures that only Th that are specific for the antigen contact the B cells. There are also interactions between other molecules on the surface of the B and Th. One interaction involves a molecule called CD40 on the B cell binding to a molecule called CD154 on the T cell. Another important interaction appears to be between a protein called inducible co-stimulator (ICOS) on the Th, which binds to a signalling protein called B7RP on B cells. Individuals with mutations in their ICOS genes have defective antibody production. As a consequence of these cell-surface inter-

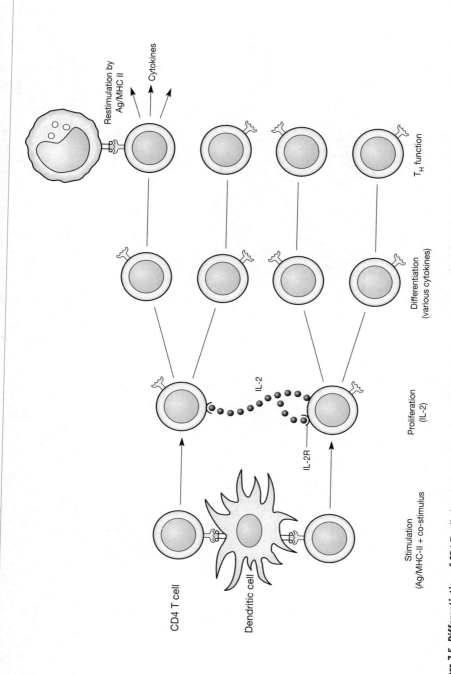

Figure 7.5 Differentiation of CD4 T cells into helper T cells. Naive CD4 T cells are stimulated by Ag/MHC-II and, if they also receive the co-stimulus, they express receptors for IL-2 (IL-2R) and secrete IL-2, which causes proliferation of the activated CD4 T cells. After proliferating, the CD4 T cells differentiate into effector Th, which, when restimulated by the same Ag/MHC-II, secrete cytokines that affect the function of other cells.

Figure 7.6 Early events in antibody production. (a) Lymph node. Antigen stimulates B cells in follicles in the cortex ① and CD4 T cells (presented by dendritic cells) in the paracortex ②. Stimulated B cells and CD4 T cells, migrate to the border of the cortex and paracortex ③ where B cells present antigenic peptide on class II MHC to the CD4 T cells, which in turn give signals to the B cells to proliferate and differentiate ④. Some of the B cells differentiate into plasma cells, which migrate to the medullary cords and secrete antibody ⑤. A few of the B cells and some CD4 T cells enter a primary follicle ⑥ where they will form a germinal centre. (b) Spleen. Antigen stimulates B cells at the junction of the marginal zone and primary follicle ① and CD4 T cells in the PALS ②. The CD4 T cells migrate to the B-cell area ③ where CD4 T-cell/B-cell interactions take place, resulting in proliferation of B cells ④ and formation of some plasma cells in the same way as in the lymph node. The plasma cells migrate to the medullary cords and secrete antibody ⑤. Other B cells and CD4 T cells enter primary follicles, leading to the development of a germinal centre ⑥.

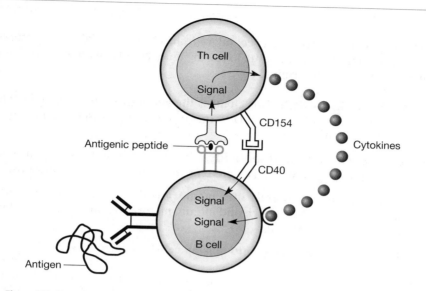

Figure 7.7 Interactions between B cells and Th. B cells bind antigen via their membrane Ig, endocytose and process it and present peptides derived from the antigen in association with class II MHC on the B-cell surface. Th cells recognising Ag/MHC-II on the B-cell surface are stimulated to secrete cytokines, which deliver signals to the B cells. Binding of CD154 on the Th cell to CD40 on the B cell delivers additional signals to B cells, which are essential for the B cells to switch antibody class.

actions and cytokine production the Th and B cells are both stimulated to proliferate and eventually form a cluster, or focus, of cells at the outer edge of the paracortex. Although the exact details of this proliferation is not known, *in vitro* B cells can be stimulated to proliferate by IL-2, IL-4 or IL-5 and it assumed that one or more of these cytokines plays a similar role *in vivo*. Some of the B cells differentiate into plasma cells secreting IgM. It is thought that other B cells undergo class switch to IgG and these can then differentiate into IgG-secreting plasma cells. Class switch is described in more detail in Section 7.4.3. The foci of IgM- and IgG-secreting plasma cells are often the main source of antibody in a primary response and reach their peak about 4 days after exposure to antigen. After 4–7 days some of the B cells and Th cells migrate to primary follicles, which is where germinal centres will form.

Antigen in the marginal zone of the spleen can be picked by the many macrophages in the region and presented to B cells in the primary follicles (Figure 7.6b). The B cells migrate to the border of the PALS and primary follicle where they meet Th cells that have also migrated there. The same molecular interactions occur between the B cells and Th as seen in the lymph node resulting in the rapid division of the cells, the formation of foci and the production of IgM and IgG. As in the lymph node, some of the B cells and Th cells eventually migrate to primary follicles and initiate the formation of germinal centres.

7.4 Formation of germinal centres (4–14 days after antigen)

Germinal centres, also called secondary follicles, are specialised structures that form within lymphoid tissue following an encounter with antigen. Four main events occur in germinal centres:

- **Antibody class switch.** This changes the antibody on the B lymphocyte cell surface from IgM and IgD to one of the other classes of antibody – IgG, IgA or IgE.

- **Affinity maturation of antibody.** This results in antibody with high affinity for antigen, which is more effective at dealing with pathogens.

- **Differentiation of B cells into memory cells.** These are B cells that have undergone class switch and affinity maturation but have not differentiated into plasma cells.

- **Differentiation of B cells into plasma cells.**

Two of these events, affinity maturation and the production of memory B cells, occur only in germinal centres. The process of germinal centre formation is the same in all organised lymphoid tissue and is as follows (Figure 7.8).

7.4.1 Initiation of germinal centres

One, or a few, B cells together with some Th from a foci of antibody-producing cells enter a primary follicle. Primary follicles are specialised structures containing a unique type of antigen presenting cell called a **follicular dendritic cell**. Follicular dendritic cells form a dendritic network throughout the follicle and are very good at retaining antibody–antigen complexes on their surface and presenting the antigen to B cells. The antibody for the immune complex is produced initially by the plasma cells in the extra-follicular foci.

The migration of B cells and Th cells into follicles is promoted by the chemokine B lymphocyte chemoattractant, which is produced by follicular dendritic cells and other stromal cells in the B-cell area and binds to the chemokine receptor CXCR5 expressed on B cells and Th.

7.4.2 Affinity maturation

The B cells that have entered the primary follicle down-regulate their Ig membrane receptors and undergo extensive proliferation. At this stage they are called centroblasts. During this period of proliferation the B cells undergo a process called **affinity maturation**. It is important to produce antibody with high affinity for antigen because high-affinity antibody works better. Affinity maturation occurs in the following way.

Plasma cell

Memory B cell

Th cell

Apical light zone –
differentiation into
plasma or memory cell

Centrocyte

FDC

Basal light zone –
selection by FDCs

Centroblast

Dark zone – proliferation
and somatic mutation

Figure 7.8 Germinal centre formation. B cells proliferate in the dark zone, where they are known as centroblasts. During this proliferation, somatic mutation of the B-cell Ig takes place. The centroblasts stop dividing and enter the basal light zone, where they are known as centrocytes. The centrocytes encounter antigen being presented by follicular dendritic cells and those centrocytes whose Ig binds antigen are rescued from cell death. Centrocytes whose Ig does not bind antigen die by apoptosis. Surviving centrocytes enter the apical light zone, where they present antigen on their class II MHC to Th cells. B cells presenting antigen that is recognised by the Th cells receive a survival signal from the Th cell and differentiate into plasma cells or memory cells. B cells presenting antigen for which there are no Th die. Note that although centroblasts and centrocytes are given special names, they are in fact B cells at special stages of differentiation.

As the centroblasts in the germinal centre divide (Figure 7.8) special molecular mechanisms cause hyper-mutation of the H and L chain variable genes. Because this is happening in non-germ-line cells (germ-line cells are sperm and ova) it is called somatic mutation (Figure 7.9). The mutations in the Ig genes lead to changes in the nucleotide sequence and some of these changes will alter the amino-acid sequence of the hyper-variable regions of the antibody molecule and therefore change the conformation of the anti-gen binding site. Since the mutations are random, the change in the conformation of the antigen binding site could either increase or decrease the affinity of the antibody for antigen or have no effect on the affinity.

Figure 7.9 Somatic mutation of Ig genes. During somatic mutation the nucleotide sequence of V genes of the Ig H- and L-chains changes. The black lines in the lower figure indicate sites where the nucleotide sequence has changed from the original germ-line V-gene sequence. Note that somatic mutation occurs in parts of the V gene coding for both CDR1 and CDR2 and the framework (non-CDR) parts of the V genes. There is, however, no somatic mutation in the C regions of the genes.

Since we want only the B cells that have mutated their Ig genes to produce high-affinity Ig, it is necessary to select for these high-affinity cells and this occurs as follows.

The centroblasts stop dividing and re-express their membrane Ig; they are now called **centrocytes** (Figure 7.8). The centrocytes must now recognise antigen on the follicular dendritic cells to receive a survival signal. Centrocytes whose surface Ig has high enough affinity for antigen will successfully bind to the antigen on the follicular dendritic cell and receive a survival signal. Centrocytes with lower affinity for antigen will not be able to bind to the antigen and do not receive the survival signal; these cells die by apoptosis and are engulfed by macrophages (Figure 7.10). As the immune response progresses, antibody will begin to remove antigen and the amount of antigen available for continued stimulation of B cells will fall. Once the amount of antigen becomes limiting, the B cells will begin to compete with each other for antigen. In this competition, B cells that have mutated their antibody to a higher affinity will win the competition and be the only ones that will continue to be stimulated and undergo cell division. This will lead to the generation of many B cells with high affinity for antigen and the death of low-affinity B cells. This process of affinity maturation is extremely effective and may result in a 10,000- to 100,000-fold increase in antibody affinity as the immune response progresses.

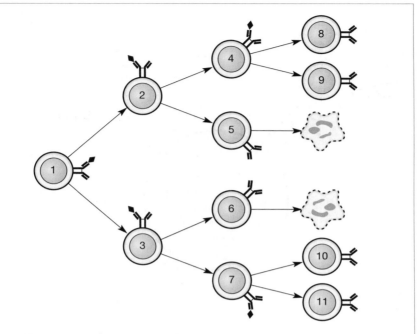

Figure 7.10 B cells die unless stimulated by antigen. In this example, B cell number 1 binds antigen and divides to produce daughter cells 2 and 3. Cells 2 and 3 both bind antigen and divide again to produce cells 4, 5, 6 and 7. Cells 4 and 7 bind antigen and produce daughter cells 8, 9, 10 and 11. Cells 5 and 6 do not bind antigen and die by apoptosis.

Centrocytes and centroblasts are found in different regions of the germinal centre (Figure 7.8). Centroblasts are found in the dark zone and centrocytes in the basal light zone. The basal light zone contains many follicular dendritic cells for affinity maturation and macrophages to mop up the apoptotic B cells. It is thought that some cells can accumulate up to 20 mutations in their Ig genes, although other cells undergo less somatic mutation.

7.4.3 Class switch

During the centroblast/centrocyte stage B cells undergo another differentiation process called the antibody **class switch** (Figure 7.11). Using special molecular mechanisms, the B cell is able to change the heavy chain constant regions of its antibody from μ and δ to γ, α or ε while keeping the same heavy chain variable region and light chain (Box 7.2). Therefore the class switch does not change the antigen specificity of the antibody, but just the class of Ig.

Class switch is controlled by the Th cells and cytokines. As mentioned above the interaction between B cells and Th involves both cell contact and cytokines (Figure 7.7). The direct cell–cell interactions between the B cells

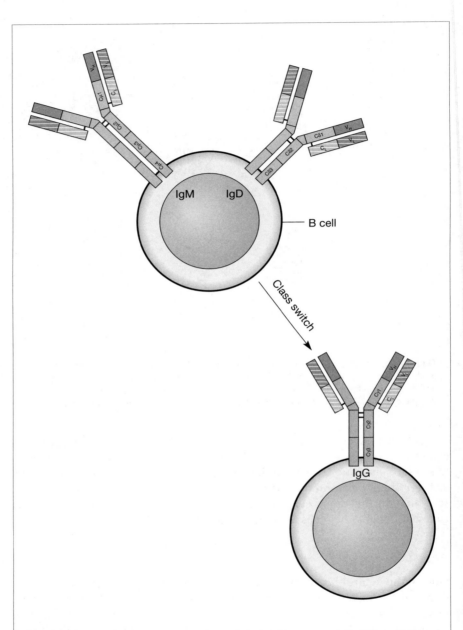

Figure 7.11 B-cell class switch. During class switch, B cells stop expressing IgM and IgD on their cell surface and express another class of antibody, in this example IgG. Note that the V_H and the L-chain stay the same; only the C regions of the H-chain change.

and Th are thought to be the same as described above. B cells express class II MHC on their cell surface and present antigen to the Th that stimulates the Th to produce cytokines. CD40 on the B cell binds to CD154 on the T cell. This interaction between CD40 and CD154 is essential for the class switch and without it only IgM is produced. The importance of this interaction in humans was demonstrated in a congenital (genetically determined) immunodeficiency disease known as hyper-IgM syndrome. Individuals with this syndrome have high levels of circulating IgM but little or no IgG, IgA or IgE. It has been demonstrated that these individuals have mutations in their CD154 gene so that they either do not make CD154 or they make a mutated form of the protein that binds poorly, if at all, to CD40. Because they cannot switch antibody class, affected individuals give especially poor secondary antibody responses and are prone to certain types of infections. Additionally ICOS on the Th interacts with B7RP on the B cells. Although the CD154/CD40 and ICOS/B7RP interactions are essential for the class switch to occur, they do not appear to influence the class of Ig that a B cell switches to. This appears to be influenced more by cytokines produced by the Th when it interacts with the B cell.

BOX 7.2: MOLECULAR ASPECTS OF CLASS SWITCH

The Ig genes contain special regions called switch regions, which allow the DNA to recombine so that the V gene moves from being immediately upstream of the Cµ gene to being upstream of another C-region gene; in the diagram the V region switches to being upstream of Cγ1.

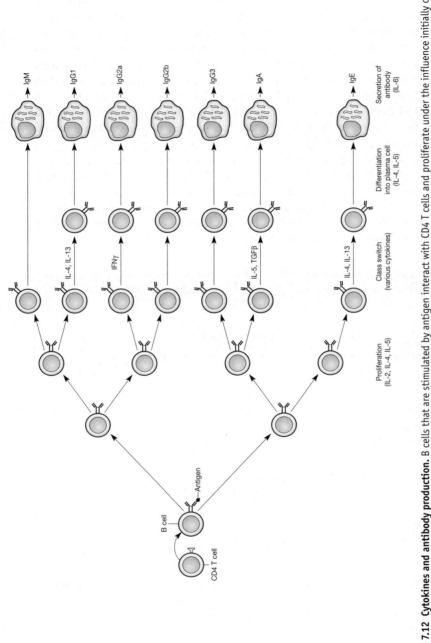

Figure 7.12 Cytokines and antibody production. B cells that are stimulated by antigen interact with CD4 T cells and proliferate under the influence initially of IL-2 and later IL-4 and IL-5. After proliferating, the B cells undergo class switch controlled by various cytokines secreted by Th cells. Finally, differentiation into plasma cells is influenced by IL-4 and IL-5 and antibody secretion is stimulated by IL-6. Some cytokines, e.g. IL-6, may be secreted by cells other than Th. In reality not all classes of antibody are produced in response to a single pathogen.

Cytokines affect the proliferation, differentiation and class switching of B cells and also the level of antibody production by plasma cells. Although there are some features in common between humans and mice regarding cytokines and the production of IgA and IgE, the control of IgG subclasses shows considerable differences in the two species. IL-4 promotes the switch to IgE in both species. In mice IL-13 can substitute for IL-4 and the situation is that either IL-4 or IL-13 is essential for IgE production. Transforming growth factor-β (TGFβ) promotes IgA production in both species. The situation with regards to IgG subclasses is not so clear cut, especially in humans. In the mouse IL-4 and IL-13 promote IgG1 production and IFN-γ increases IgG2a. However, IFN-γ appears to have no direct effect on B cell switching in humans.

Table 7.1 and Figure 7.12 show the cytokines involved in plasma cell generation in the mouse; although the exact cytokines are less well known in humans, it is very likely that presently unidentified factors will control B cell differentiation in a similar fashion to that seen in mice.

7.4.4 Differentiation into plasma or memory B cells and antibody production

The final stage of B cell differentiation is the differentiation of B cells into plasma cells or memory B cells (Figure 7.8). Plasma cells secrete large amounts of antibody (Figure 7.13) to deal with the current threat while memory B cells provide protection against future infection. Memory B cells are long-lived and can survive long after an initial infection has been eliminated. Memory B cells have undergone affinity maturation and class switch but retain surface expression of the class of Ig they have switched to. They enter the recirculating lymphocyte pool and migrate through lymph nodes, spleen and some other lymphoid tissue. If they encounter antigen in the future they are able to very rapidly develop into plasma cells secreting high-affinity IgG, IgA or IgE because they do not need to undergo affinity maturation or class switch again. Therefore upon second encounter with an antigen you get a faster and bigger IgG, IgA or IgE antibody response (see Chapter 10).

Table 7.1 Influence of cytokines on antibody production in mice

Antibody class	Cytokines promoting switching/production
IgG1	IL-4, IL-13
IgG2a	IFNγ
IgG2b	TGFβ
IgG3	IFNγ
IgA	IL-5, TGFβ
IgE	IL-4, IL-13

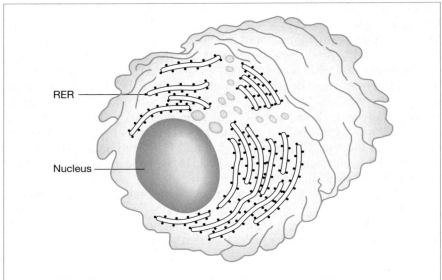

Figure 7.13 Diagram of plasma cell. Note the extensive RER typical of a cell synthesising large amount of protein for secretion. In this case the secreted protein is antibody.

Each B cell in the apical light zone can differentiate into either a plasma cell or a memory B cell (Figure 7.8). The factors that affect whether a B cell becomes a plasma cell or a memory cell are not fully understood. There is good evidence that if the B cell interacts with a Th, the binding of CD154 on the Th to CD40 on the B cell stimulates the B cell to become a memory B cell and both Th and CD154/CD40 interactions are essential for memory B cell generation. It is not clear whether other signals stimulate the cells to become plasma cells or whether the B cells automatically differentiate into plasma cells unless they are restimulated through CD40. Both IL-4 and IL-5 promote the differentiation of B cells into plasma cells and IL-6 increases antibody production by plasma cells.

Plasma cells can be short- or long-lived. Some stay in the lymphoid tissue where they are produced, the medulla of the lymph node, the sinusoids of the splenic red pulp or mucosal associated lymphoid tissue. Plasma cells in the spleen and lymph node tend to be relatively short-lived and secrete antibody for a few weeks. Other plasma cells migrate to the bone marrow where they are maintained by interactions with the bone marrow stromal cells. These plasma cells may secrete antibody for many months or years. The maintenance of antibody production can be truly remarkable; a vaccine for the yellow fever virus was tested in a trial in the USA and some individuals who had received the vaccine still had antibody against the virus 70 years later with no deliberate reinfection.

Figure 7.14 summarises the different fates that can befall B cells in germinal centres during the antibody production process.

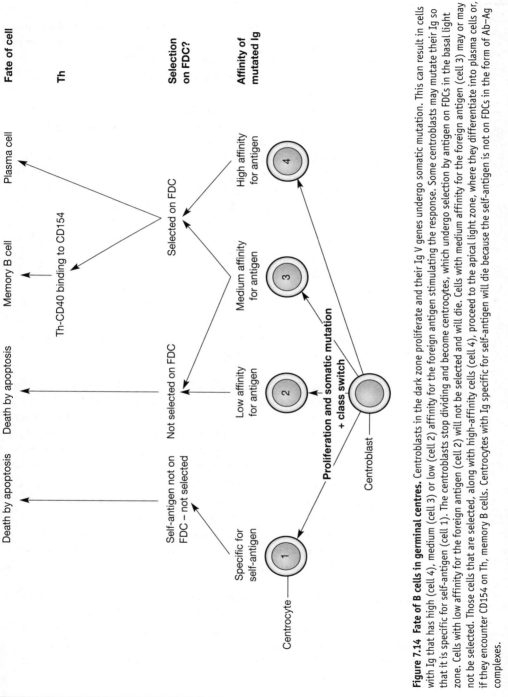

Figure 7.14 Fate of B cells in germinal centres. Centroblasts in the dark zone proliferate and their Ig V genes undergo somatic mutation. This can result in cells with Ig that has high (cell 4), medium (cell 3) or low (cell 2) affinity for the foreign antigen stimulating the response. Some centroblasts may mutate their Ig so that it is specific for self-antigen (cell 1). The centroblasts stop dividing and become centrocytes, which undergo selection by antigen on FDCs in the basal light zone. Cells with low affinity for the foreign antigen (cell 2) will not be selected and will die. Cells with medium affinity for the foreign antigen (cell 3) may or may not be selected. Those cells that are selected, along with high-affinity cells (cell 4), proceed to the apical light zone, where they differentiate into plasma cells or, if they encounter CD154 on Th, memory B cells. Centrocytes with Ig specific for self-antigen will die because the self-antigen is not on FDCs in the form of Ab–Ag complexes.

7.4.5 The combination of class switch and affinity maturation significantly reduces the time taken to make antibody

Class switch and affinity maturation appear to be other examples of the immune system being unnecessarily complicated. If you need to produce different classes of antibody with high affinity for antigen, surely the easiest way is to produce B lymphocytes already expressing different classes of high-affinity antibody and just select those B lymphocytes with the class of antibody required to become plasma cells secreting the same high-affinity antibody of the required class. In fact, having class switch and affinity maturation is more efficient for the following reasons.

Remember that each B cell has only one antigen specificity and that different B cells have different antigen specificities. The immune system must provide B cells, which as a population have as many antigen specificities as possible. It is not possible to escape this responsibility because that would leave you open to infection by pathogens whose antigens you did not have B cells specific for. However, your body only has room for a fixed number of B cells. Therefore you can only have a certain number of B cells that are specific for any one antigen. The more B cells you have that are specific for an antigen, the quicker you will generate enough plasma cells to produce the amount of antibody needed to protect you against the pathogen. Class switch and affinity maturation maximise the number of B cells that can be initially stimulated by a single antigen to become high-affinity antibody producing plasma cells in the following ways.

Class switch If class switch did not occur, you would need a separate B cell with the same antigen specificity for each class of antibody. Since there are eight classes of antibody in humans you would need eight times as many B cells to cover the same range of antigen specificities. Because you cannot reduce the number of antigen specificities covered by your B cell population as a whole, and you cannot increase the total number of B cells that your body can accommodate, the only solution to this problem would be to have eight times fewer B cells specific for any particular antigen. This would not be a problem if you always made all eight different classes of antibody. However in most infections a few classes of antibody predominate and you do not make some classes at all. Therefore B cells of the class you do not want are wasted (Figure 7.15).

Affinity maturation Affinity maturation enables more B cells to be initially stimulated by antigen in the following way. B cells must bind antigen with a certain affinity to be stimulated to eventually become plasma cells. For any particular antigen some B cells will have Ig on their surface with high affinity for the antigen but there will be many many more B cells whose Ig binds the antigen with much lower affinity. It is difficult to estimate exactly how many more low-affinity B cells there would be compared with high-affinity cells but an estimate of 1000-fold is realistic. If affinity maturation did not occur, the only way to make high-affinity antibody would be to select the very small number of high-affinity B cells for clonal expansion. This would require more rounds of cell division to generate the required number of plasma cells. By having affinity maturation you can also select the low-affinity B cells for clonal expansion and allow them to affinity-mature, thereby initially stimulating 1000 more antigen-specific B cells.

The combination of affinity maturation and class switch means that when a new antigen comes along you may have 8000 times as many B cells to select for clonal expansion compared with the situation where affinity maturation and class switch did not occur. In other words, **without** class switch and affinity maturation you have to generate 8000 B cells from a single B cell to be in the same starting position as you are **with** class switch and affinity maturation (Figure 7.15). Generating 8000 cells from one cell involves 13 cell divisions. Assuming that B cells can divide once every 10 hours, 13 generations would take 130 hours, or 5.5 days. Even allowing that affinity maturation is not 100% efficient and that you may want to make more than one class of antibody (but not all eight classes), you still save many days in generating antibody by having class switch and affinity maturation. Given the rate of multiplication of some pathogens, these days may be crucial to whether you survive an infection or not and from an evolutionary point of view would be a powerful selection force for class switch and affinity maturation.

7.5 MALT and the production of IgA

MALT contains many lymphoid follicles, which serve the same purpose as those seen in the spleen and lymph nodes. As might be expected, in areas that are exposed to many environmental antigens, most of the follicles contain germinal centres where plasma cells and memory B cells are generated. The area surrounding the follicles is rich in T cells, which will provide the Th for antibody generation.

The epithelium overlying the follicles in the gut is specialised to enable it to transport antigen into the lymphoid follicle from the lumen of the intestine. The epithelium above lymphoid follicles is flattened and does not contain villi even in mucosa, such as the ileum, which normally have many villi (see Figure 7.16). Special epithelial cells called **M cells** overly the follicles. The internal side of the M cells contains deep invaginations that surround clusters of B cells, T cells, macrophages and dendritic cells. The

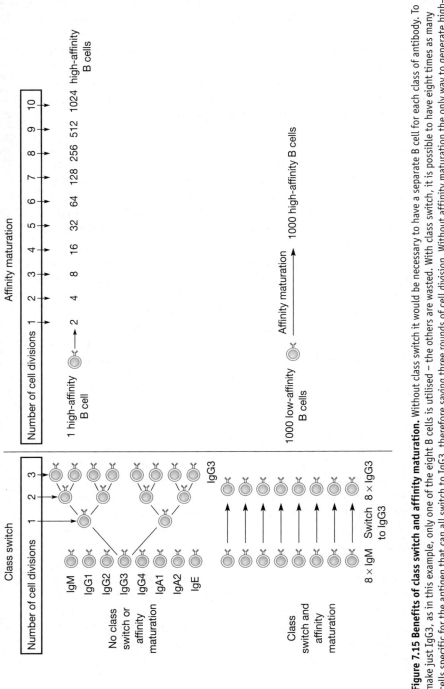

Figure 7.15 Benefits of class switch and affinity maturation. Without class switch it would be necessary to have a separate B cell for each class of antibody. To make just IgG3, as in this example, only one of the eight B cells is utilised – the others are wasted. With class switch, it is possible to have eight times as many cells specific for the antigen that can all switch to IgG3, therefore saving three rounds of cell division. Without affinity maturation the only way to generate high-affinity antibody would be to select B cells with high affinity for antigen – these would be quite rare. With affinity maturation it is possible to select low-affinity B cells, which may be present at 1000-fold higher frequency than the high-affinity cells – this saves ten rounds of cell division. Overall, if they were 100% efficient, the combination of class switch and affinity maturation would save 13 rounds of cell division, which would take about 5.5 days. Even if the processes were only 5% efficient, they would save between eight and nine divisions, or about 4 days.

Figure 7.16 Diagram of mucosal lymphoid tissue. Specialised epithelial cells called M cells have a flattened surface and are specialised for capturing antigen in the lumen and transporting it to the mucosal lymphoid tissue where antigen can stimulate B cells and be processed by dendritic cells for presentation to CD4 T cells.

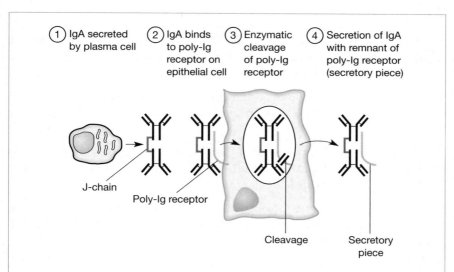

Figure 7.17 Secretion of IgA. IgA secreted by plasma cells in the mucosa binds to a receptor, called the poly-Ig receptor on mucosal epithelial cells. The poly-Ig receptor, together with the bound IgA, is internalised by the epithelial cells and the poly-Ig receptor is cleaved. Finally the IgA is secreted into the lumen with part of the poly-Ig receptor, now known as the secretory piece, attached.

M cells endocytose antigen present in the mucosal lumen and the endocytic vesicle crosses the M cell before fusing with the basal cell membrane, thereby releasing the endocytosed antigen into the invagination containing the lymphocytes.

The antigen can then be taken up by dendritic cells for presentation to T cells or recognised by B cells in the follicles. The sequence of events that follows leading to the formation of germinal centres and production of plasma and memory B cells is the same as that occurring in lymph nodes or the spleen. Activated B cell become committed to IgA, leave the follicles in the Peyer's patches and enter lymphatic vessels, eventually reaching the blood. The vascular system delivers the B cells to the lamina propria at any mucosal site in the body. The B cell now becomes an IgA secreting plasma cell. In this way the IgA-secreting cells can be distributed far from the site of original antigenic stimulation and provide protection against infection by that pathogen at all mucosal sites.

The IgA produced by the plasma cells in the lamina propria is taken up by the epithelial cells of the mucosa. The epithelial cells add the secretory piece to the IgA before releasing the IgA into the lumen (see Figure 7.17). The secretory piece inhibits the degradation of the IgA by enzymes in the mucosal secretions.

7.6 Summary

- Antigen entering tissue can enter the MALT or it can enter lymphatic ducts draining the tissue site, where it will go to the local lymph nodes. Antigen in the blood goes to the spleen.

- Antigen is taken up by dendritic cells in tissue or lymphoid tissue, and dendritic cells present antigen to, and activate, CD4 T cells to become Th cells.

- Th cells migrate to the edge of follicles, where they encounter B cells that have been stimulated by antigen. The B cells differentiate into antibody producing extra-follicular foci.

- Some B cells migrate into primary follicles accompanied by Th cells, resulting in germinal centre formation. Here plasma cells or memory B cells are produced.

- In the germinal centre B cells undergo two unique events. They switch the class of antibody they express on their cell surface from IgM and IgD to IgG, IgA or IgE, a process known as class switch. The B cells also randomly mutate their Ig V genes, therefore altering the nucleotide, and hence amino acid, sequence of the variable region. B cells whose mutated Ig has higher affinity for antigen compete favourably for antigen and continue to be stimulated by antigen. This results in selection of higher-affinity B cells – a process known as affinity maturation.

- After class switch and affinity maturation, B cells complete their differentiation and become either plasma cells, which produce large amounts of antibody, or memory B cells. Memory B cells are long-lived and able to respond rapidly to a second exposure to the same antigen. This gives a bigger, better and longer response to the antigen and is known as a secondary antibody response.

CHAPTER EIGHT

Effector mechanisms: dealing with pathogens *in vivo*

(1) Antibody-mediated responses

Learning objectives

To know about the different ways the specific immune response aids in neutralising or eliminating pathogens. To understand the different ways in which antibody provides protection from infectious agents. To learn about the components and biology of the complement system.

Key topics

- Antibody-mediated effector mechanisms
  - Neutralisation
  - Agglutination
  - Opsonisation
  - Antibody-dependent cell-mediated cytotoxicity (ADCC)
- Killing by phagocytes
- Complement
  - Components of the complement system
  - Three pathways of complement activation
  - Biology of the complement system

8.1 Humoral and cell-mediated immunity

The ways in which effector responses can protect the individual are by first of all neutralising any immediate threat posed by the pathogen (e.g. toxins produced by the pathogen) and then bringing about the destruction and/or elimination of the pathogenic organism so that it no longer has the potential to cause disease. The three types of specific effector response were mentioned in Chapter 6; these are antibody responses, cytotoxic CD8 T cell responses and delayed-type hypersensitivity responses. Antibody responses have also been called humoral immunity because, historically, the immunity could be transferred with 'humour'; that is, liquid (this was serum and not funny jokes). Cell-mediated immunity was so-called because it could only be transferred with cells and not with humour – it is now clear that cell-mediated immunity is referring to immune responses involving CD8 T cells or delayed-type hypersensitivity responses, neither of which involve antibody.

8.2 Antibody-mediated effector responses

Antibody can protect against infectious agents in a number of ways, many of which involve interaction with components of the innate immune system. These can be summarised as:

- neutralisation;
- agglutination;
- opsonisation;
- activation of complement;
- antibody-dependent cell-mediated cytotoxicity.

As described in Chapter 3, there are many different classes and subclasses of antibody. Each class and subclass of Ig has a different Fc portion of the IgH chain and these different Fc portions endow the antibodies with different functions.

8.3 Neutralisation by antibody

This is the simplest way in which antibody can act as an effector molecule and a variety of pathogens or their products can be neutralised by antibody (Figure 8.1).

8.3.1 Toxins

Antibody can neutralise toxins simply by binding to them and thereby inhibiting their action. This is particularly important with diseases such as tetanus, diphtheria and botulism (a severe type of food poisoning), where the pathology is due totally to production of powerful toxins.

8.3.2 Virus

For a virus to infect a host cell, specific molecules on the viral surface must bind to molecules on the surface of the host cell (see Box 8.1). Antibody

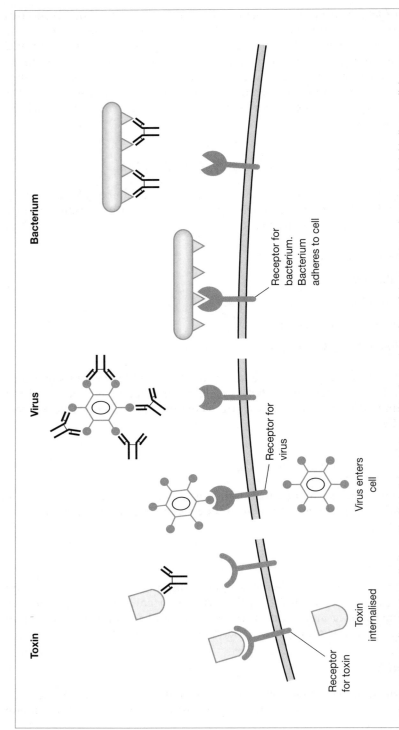

Figure 8.1 Neutralisation by antibody. Antibody binding to toxins or molecules on the surfaces of viruses or bacteria prevents their binding to cellular receptors. This prevents internalisation of toxins or viruses and prevents adhesion of bacteria to cell surface.

Viruses have to infect host cells in order to replicate. The first stage in infecting the cell is the attachment of the virus to a molecule on the host cell. In some cases, such as 'flu or cold viruses, the molecules that the virus attaches to are expressed on many cell types. In other cases the molecule has a very restricted cellular distribution, which determines the tissue specificity (tropism) of the virus.

Virus	Disease	Host cell attachment molecule
Influenza	'Flu	Sialic acid on glycoproteins
Rhinovirus	Cold	ICAMs
Rabies virus	Rabies	ACh-R
HIV	AIDS	CD4 (+CCR5 or CXCR-4)
Epstein–Barr virus	Infectious mononucleosis	C3 receptor 2 on B cells
Herpes simplex 1 virus	Cold sores	Fibroblast growth factor receptor (FGF-R)
Rotavirus	Infantile diarrhoea	β-Adrenergic receptor

Bacterial adhesion is often of a much less cell-specific nature because the molecule on the host cell is found in many cell types. A very important role for bacterial adhesion is to allow the bacteria to colonise mucosal surfaces in the GI, respiratory and GU tracts.

Gram-negative bacteria (e.g. E. coli, Vibrio cholera and Neisseria gonorrhoeae) can produce pili – small hair-like extensions from the bacterial cell surface that are smaller than flagella. These pili can enable the bacteria to adhere to epithelial cells. Many bacteria have pili that bind to α-mannosides.

Both Gram-positive and Gram-negative bacteria have non-pili-associated adhesion molecules. Bordatella pertusis, the cause of whooping cough, produces a number of adhesins. One, called filamentous haemagglutinin (FHA), binds to ciliated cells. It also contains the RGD motif (arginine-glycine-aspartate) that enables it to bind to CR3 (CD11a/CD18 integrin) on macrophages. Neisseria gonorrhoeae makes proteins called opacity proteins that are involved in binding of the bacteria to epithelial cells. Staphylococcus aureus produces a protein that can bind to fibronectin on epithelial cell surfaces.

can bind to the viral receptor molecule and stop it binding to the cell, thereby preventing infection of the cell. Antibody present in mucosal secretions, particularly IgA, can prevent viral entry into the body and stop the infection from being established. Many viruses spread through the bloodstream to their target organs and during this extracellular phase they are exposed to circulating antibody that can bind to the virus and prevent it from infecting the target cells.

Antibody can also prevent the virus from replicating after it has entered the cell. In ways that are not understood, antibody prevents the virus from uncoating, a necessary prerequisite to replication; this has been suggested to occur for measles and influenza viruses. Other antibodies inhibit viral replication in ways that are poorly understood.

8.3.3 Bacteria

In addition to neutralising bacterial toxins, antibody can also affect bacterial adherence to cells and inhibit bacterial metabolism. Mucosal IgA can bind cholera bacteria and prevent their adherence to the intestinal epithelium. Other antibodies have been found to inhibit bacterial metabolism by binding to pores in the bacterial cell wall that are involved in nutrient transport.

8.4 Agglutination

Because antibodies are multivalent they are able to bind to more than one microbial particle and can form complexes of microbe and antibody (Figure 8.2), a process known as agglutination. Agglutination can limit the spread of pathogens by retaining them in clumps; larger complexes of antigen are more likely to be phagocytosed and killed (see Section 8.5).

8.5 Phagocytosis and killing

Phagocytosis involves the engulfment and ingestion of a particle. Usually these particles are microorganisms, but damaged cells or tissue components can also be phagocytosed in the process of tissue cleansing. Phagocytosed microbes will normally be killed by the phagocyte, although some (e.g. mycobacteria) have developed resistance to killing by phagocytes.

Phagocytosis and killing can be divided into four stages (see Section 2.2.4):

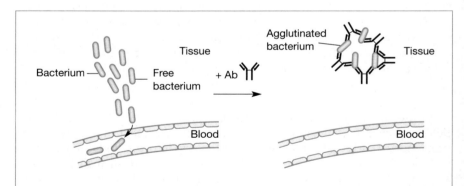

Figure 8.2 Agglutination. Antibody binds to antigens on bacteria, causing them to agglutinate or clump, which inhibits the bacteria from leaving the site and also makes bacteria more susceptible to phagocytosis.

1. Recognition and attachment.
2. Ingestion.
3. Killing.
4. Degradation.

8.5.1 Recognition and attachment – antibodies can act as opsonins

The first stage of phagocytosis involves recognition of the microorganism by the phagocyte. As mentioned in Chapter 2, phagocytes can recognise microbes directly using specific receptors on their cell surface that can recognise (bind to) various molecules on microorganisms. However, many pathogens have evolved ways of avoiding direct recognition by phagocytes and the phagocytes need help in recognising pathogens. Opsonins are molecules that bind to pathogens and to phagocytes and promote phagocytosis (see Chapter 2). Some opsonins belong to the innate immune system but are limited in which pathogens they can recognise. Because they are specific for antigens on the pathogen, antibodies can be very effective opsonins and recognise pathogens that opsonins of the innate immune system cannot.

Antibodies can act directly as opsonins by binding to the pathogen using their Fab antigen binding sites and then binding to receptors on the phagocyte that are specific for the Fc part of the antibody molecule (Figure 8.3). The receptor on the phagocyte that binds the Fc portion of the antibody is called, not surprisingly, an Fc receptor, or FcR. There are a number of different FcRs that bind different classes of antibody and their main properties are listed in Table 8.1. It should be noted that not all the Fc receptors listed in Table 8.1 are involved in phagocytosis. Fcγ RI and Fcγ RIII promote phagocytosis. Fcγ RII is an inhibitory receptor and inhibits macrophage and other cell functions. The IgE- and IgA-binding FcRs are involved in specialised activities of IgA and IgE that do not include phagocytosis. When antibody binds to

Table 8.1 Fc receptors

Receptor	Ig bound	Distribution
Fcγ RI	IgG1, 3	Macrophages, dendritic cells Inducible on neutrophils, eosinophils
Fcγ RII	IgG1, 3	Different forms with different intracellular regions expressed on all leucocytes except T cells
Fcγ RIII	IgG1, 3	Monocytes, neutrophils, T cells, NK cells, eosinophils, mast cells, FDCs
Fc_ϵ RI	IgE	Mast cells, basophils, FDCs Inducible on eosinophils
Fc_ϵ RII	IgE	B cells, macrophages, eosinophils
FcaRI	IgA1, 2	Macrophages, neutrophils Inducible on eosinophils

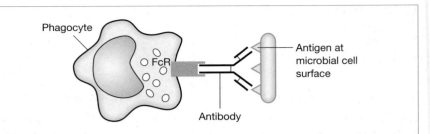

Figure 8.3 Antibody as an opsonin. Antibody binds to antigen on microbial surfaces via the antigen-binding site. Phagocytes have FcRs that bind to the Fc part of the antibody molecule and the antibody/microbe complex is phagocytosed.

antigen, the Fc portion of the antibody undergoes a conformational change that enables it to bind to FcR on phagocytes. The binding of antibody to FcRs triggers the phagocyte to phagocytose and, it is hoped, to kill the ingested pathogen. The FcRs therefore play an active role in triggering phagocytosis and do not simply bind the antibody–antigen complex. Opsonisation by antibody can also be promoted by complement (see Section 8.6).

8.5.2 Ingestion

Once phagocytosis has been triggered the phagocyte extends its membrane around the particle, forming structures called pseudopodia. Eventually the particle becomes completely surrounded by the phagocytic cell and is engulfed in a phagocytic vacuole. This completes the process of ingestion and the phagocyte will now normally proceed to kill the ingested organism.

8.5.3 Killing

When the microbe is inside a phagocytic vacuole the phagocyte will attempt to kill it. Phagocytes have a number of mechanisms for killing ingested microorganisms. Some of these can be performed within the vacuole by mechanisms that are activated by the formation of the vacuole and by microbial products. These do not require fusion of the phagosome with a lysosome and hence are termed lysosome-independent. However, not all microorganisms are killed by these lysosome-independent mechanisms. The phagosome may, and usually does, fuse with a lysosome, resulting in exposure of the phagocytosed microorganism to lysosomal-dependent killing mechanisms (Figure 8.4).

Lysosome-independent killing mechanisms
There are two main microbicidal pathways that can occur without lysosomal fusion with the phagosome. These are the generation of **oxygen radicals** and the production of **nitric oxide**.

- **Oxygen radicals.** Oxygen radicals are highly reactive chemicals with bactericidal activity. They can bind to and damage a variety of microbial products such as membranes, proteins and DNA, leading to killing

Plate 1. Elephantiasis. Obstruction of lymphatic vessels in chronic filiarisis caused by the nematode worm *Wucheria bancrofti* can lead to massive oedema and other complications, resulting in elephantiasis. (Source: Courtesy of Richard Suswillo, Department of Infectious Diseases, School of Medicine, Imperial College, London, UK.)

(a)

(b)

(c)

Plate 2. Cells of the phagocyte lineage. (a) Monocytes circulate in the bloodstream, where they are not phagocytic. Monocytes leave the bloodstream and enter sites of inflammation, where, under the influence of inflammatory mediators, they differentiate into inflammatory macrophages (H&E × 1000). (b) Tissue macrophages are weakly phagocytic unless stimulated by microbial products or inflammatory mediators. The macrophages (M) have phago-cytosed black carbon particles (× 400). (c) Neutrophils have a characteristic multi-lobed nucleus. They are usually the first cells to be recruited from the bloodstream to sites of inflammation. Once they have left the bloodstream they perform a 'kamikaze' role, rapidly phagocytosing and killing bacteria before dying themselves in a few hours. The mixture of dead neutrophils and bacteria products make up pus (H&E × 800). (Source: Courtesy of Mike Mahon, John S. Dixon and Philip F. Harris, University of Manchester.)

(a)

(b)

Plate 3. Mast cells and basophils. These cells have many similarities and are characterised by numerous large granules. (a) Basophils are white blood cells and can enter some inflammatory sites (H&E × 800). (b) Mast cells are found in connective and mucosal tissue. This picture is of the bladder wall and the mast cell granules are stained red. The mast cell third from the right is partially degranulated (pararosanaline alpha napththyl stain × 400). Despite their similarities there is no convincing evidence that mast cells are derived from basophils. (Source: Courtesy of Mike Mahon, John S. Dixon and Philip F. Harris, University of Manchester.)

Plate 4. Lymphocyte and plasma cell. (a) The lymphocyte consists of a darkly staining nucleus with a narrow layer of cytoplasm (H&E × 1000). (b) Plasma cells are bigger than lymphocytes and have much more cytoplasm (H&E × 4000). (Source: Courtesy of Mike Mahon, John S. Dixon and Philip F. Harris, University of Manchester.)

(a)

(b)

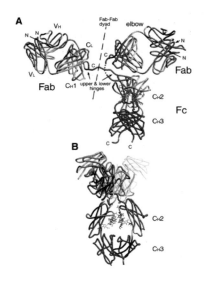

Plate 5. Antibody structure. This plate shows the structure of an antibody directed against the gp120 envelope protein of HIV. The antibody shows a globular Y-shaped structure with asymmetric arrangement of the variable regions due to conformation of the hinge regions. (Source: From Saphire, E.O. etal. (2001), 'Crystal structure of a neutralising human IgG against HIV-1: A template for vaccine design'. *Science*, Vol. 293, pp. 1155–59. Copyright © 2001 American Association for the Advancement of Science (AAAS). Reprinted with permission.)

Plate 6. Antibody–antigen interaction. The antibody (in stick form) is shown interacting with antigen, hen-egg lysozyme, in purple. The interface between antibody and antigen is flattish with dents and bumps formed by the amino-acid side chains. (Source: Courtesy of Dr. Annemarie Honegger and Prof. Dr. Andreas Plueckthun, Biochemisches Institut Der Universitat Zurich, Winterthurerstrasse 190, CH-8057 Zurich, Switzerland. Reprinted with permission.)

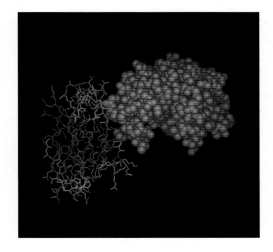

Class I MHC   Class II MHC

Plate 7. C1 I and C1 II MHC. The top figures show class I and II MHC with the peptides binding to the MHC shown in blue. The bottom figures show an overlay of class I (blue) and class II (red) MHC from the top c) and side d) showing the overall similarity in structure of the two molecules. (Source: From Stern, L.J. and Wiley, D.C., 'Antigenic peptide binding class I and class II histocompatibility proteins'. *Structure*, Vol. 2, pp. 245–51. Copyright © 1994 Elsevier. Reprinted with permission.)

Plate 8. Interaction between the T cell receptor (TcR) and antigen/class I MHC. The class I MHC molecule is at the bottom of the picture with the α-chain in green and β_2-microglobulin in yellow. The peptide being presented by class I MHC is in red. The Vα chain of the TcR is in purple and the TcR β-chain in blue; for simplicity the Cα part of the TcR is not shown. (Source: Courtesy of Dr. Annemarie Honegger and Prof. Dr. Andreas Plueckthun, Biochemisches Institut Der Universitat Zurich, Winterthurerstrasse 190, CH-8057 Zurich, Switzerland. Reprinted with permission.)

Red pulp

Sinusoid

Arteriole
Germinal centre
Follicle

(a)

Plate 9. (a) Histological section of part of the spleen (H&E × 20). (b) Section of part of a lymph node (H&E × 15). (c) High endothelial venules in a lymph node, indicated by arrows. Note that the endothelial cells are cuboidal in shape, rather than the usual flat shape of endothelial cells, giving rise to the term 'high endothelium' (H&E × 200). (Source: Courtesy of Mike Mahon, John S. Dixon and Philip F. Harris, University of Manchester.)

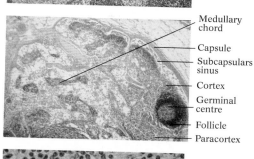

Medullary chord

Capsule
Subcapsulars sinus
Cortex
Germinal centre
Follicle
Paracortex

(b)

(c)

Peyer's patch

Germinal centre

(a)

Plate 10. Histological appearance of mucosal associated lymphoid tissue (MALT). (a) Section of ileum with Peyer's patches. Note the flattened appearance of the epithelium above the Peyer's patches (H&E × 10). (b) Section of large intestine with lymphoid nodule (H&E × 10). (Source: Courtesy of Mike Mahon, John S. Dixon and Philip F. Harris, University of Manchester.)

(b)

Lymphoid follicle

(a)

(b)

Plate 11. Langerhans cells and dendritic cells. (a) The bright green cells are Langerhans cells in the skin stained for immunofluorescence (with antibodies to CD1a). The dendritic processes of the Langerhans cells form a web that enables the cells to take up antigen in the skin (× 100). (b) Confocal image of dendritic cells (× 400). ((b) (Source: Courtesy of Dr. Pedram Hamrah and Dr. Reza Dana, Schepens Eye Research Institute, Harvard Medical School, Boston, Massachusetts, USA.)

Plate 12. Histological appearance of thymus. The darker staining outer area of each lobule is the cortex and the lighter inner staining area the medulla (H&E × 15). (Source: Courtesy of Mike Mahon, John S. Dixon and Philip F. Harris, University of Manchester.)

Cortex

Medulla

(a)

(b)

(c)

Plate 13. Appearance of immunofluorescence and immunoperoxidase. (a) Section of skin from an individual with polyarteritis nodosa, a disease in which the blood vessel walls are weakened and IgG is deposited. The IgG has been detected by incubating the tissue with fluorescent anti-human Ig (× 480). (b) Complement deposition can be detected in the same disease, polyarteritis nodosa, with fluorescent antibody to C3 (× 480). (c) Immuno-cytochemical staining of pancreas with diabetes. The section was stained with antibody to CD4 followed by peroxidase-conjugated anti-Ig and a substrate that is turned brown by the peroxidase. The section is lightly stained with haematoxylin and the brown colouration indicates the presence of many CD4 T cells in the islets (indicated by I) (× 100).

Plate 14. Infiltrating cells in a diabetic pancreas. The islets of Langerhans are indicated by the letter I and this diabetic pancreas has extensive infiltration of mononuclear cells (small dark staining cells) into the islets (H&E × 100).

(a)

Plate 16. Allergic reactions in skin and lung. (a) Wheal and flare skin reaction to applied allergen. (b) Lung of chronic asthmatic during symptom free phase. The bronchiole and alveoli are clear (van Gieson's elastica stain × 15). (c) High-power picture of the lung of a chronic asthmatic patient during an asthmatic attack. The basement membrane is thickened and there is extensive infiltration of eosinophils into the respiratory mucosa (H&E × 80).

(b)

(c)

Figure 8.4 Phagocytosis and killing. Phagocytes that have bound opsonised microbes extend their membrane round the microbe ① eventually taking the microbe into a phagocytic vacuole ② where the microbe is exposed to lysosomal-independent killing mechanisms. The phagosome may fuse with a lysosome ③, forming a phagolysosome where the microbe is exposed to lysosomal-dependent killing mechanisms in addition to the lysosomal-independent mechanisms.

of the microbe. Oxygen radicals involved in killing by phagocytes include superoxide anion (O_2^-), hydrogen peroxide (H_2O_2), singlet oxygen ($^1O_2$) and free hydroxyl radicals ($\cdot OH$). The reactions generating oxygen radicals are shown in Box 8.2.

- **Nitric oxide pathway.** The other lysosomal-independent pathway involves the generation of nitric oxide (NO), which is highly toxic to bacteria and can also inhibit viral replication. NO can bind iron and deprive bacteria of this essential growth element. NO is produced by the combination of oxygen with nitrogen derived from the amino acid L-arginine in a reaction catalysed by the enzyme nitric oxide synthase (NOS) (see Box 8.2).

Lysosomal-dependent killing mechanisms

Although lysosomal-independent mechanisms of killing exist, under normal circumstances phagosomes containing ingested microbes fuse with lysosomes to form a phagolysosome. This process exposes the contents of the phagosome to lysosomal products, which have a variety of microbicidal activities:

- **Generation of chlorine products.** Lysosomes contain an enzyme called myeloperoxidase (MPO). This catalyses the production of hypochlorous acid (HOCl) from hydrogen peroxide and chloride; the hypochlorous acid is then converted to hypochlorite (OCl^-) and chlorine (Cl_2) (see Box 8.2). As might be expected, these chlorine-containing products are toxic to many microbes (OCl^- is used as a bleach) and contribute to phagocytic killing.

- **Defensins.** These are cationic proteins that are able to form ion pores in membranes and can kill a variety of microbes, including bacteria, fungi and some viruses.

BOX 8.2: GENERATION OF BACTERICIDAL RADICALS AND NITRIC OXIDE

Oxygen radicals

The first step in this process involves the generation of the superoxide anion, O_2^-:

$$NADPH + O_2 \xrightarrow{\text{Cytochrome b-245}} NADP + O_2^-$$

The O_2^- then spontaneously reacts with other molecules, generating a number of other oxygen radicals: these are hydrogen peroxide (H_2O_2), singlet oxygen ($^1O_2$) and free hydroxyl radicals ($\cdot OH$), which are produced as follows:

$$2O_2^- + 2H^+ \longrightarrow H_2O_2 + {^1O_2}$$

$$O_2^- + H_2O_2 \longrightarrow \cdot OH + OH^- + {^1O_2}$$

Nitric oxide

The production of nitric oxide is catalysed by the enzyme nitric oxide synthase (NOS), of which there are two types: endothelial NOS (eNOS) is constitutively expressed in many cell types; inducible NOS (iNOS) is, as its name implies, induced and is the main enzyme used by phagocytes.

The reaction is as follows:

$$O_2 + \text{L-arginine} \xrightarrow{\text{NOS}} NO + \text{citrulline}$$

Generation of chlorine products

Initially hypochlorous acid (HOCL) is generated from hydrogen peroxide (H_2O_2) and chloride (Cl^-):

$$H_2O_2 + Cl^- + H^+ \longrightarrow HOCl + H_2O$$

The HOCl generates hypochlorite (OCl^-) and combines with more chloride to generate chlorine:

$$HOCl \rightleftharpoons H^+ + OCl^-$$

$$HOCl + Cl^- \longrightarrow Cl_2 + OH^-$$

- **Proteolytic enzymes.** Lysosomes contain many proteolytic enzymes that may degrade microbial products although it is not clear how important they are in attacking intact microbes. Another lysosomal product is lysozyme; as its name implies, lysozyme is able to degrade the peptidoglycan layer of Gram-positive bacteria.

An important role of the proteolytic enzymes found in lysosomes is to degrade microbial products so they can be excreted by the phagocyte. Additionally degradation of microbial proteins by macrophages can generate antigenic peptides that can be presented on the surface of the macrophage in association with class II MHC and can therefore stimulate CD4 T cells, thereby contributing towards the specific immune response against the pathogen.

8.6 Complement

Another important function of some classes of antibody is to activate the complement system. Complement is not a single molecule but a cascade of proteins whose closest parallel is the clotting system, although the functions of complement are very different from those of the clotting system. Like the clotting system, complement consists of a series of inactive precursor proteins that are activated and then activate the next protein in the sequence.

Three complement pathways exist: the classical, alternative and lectin pathways. All three pathways have a common final stage but the earlier components in the pathways differ and the ways in which the pathways are activated are also different. When describing complement components a convention is adopted because many complement components are inactive pro-enzymes. Therefore the active forms of the protein are indicated by the suffix **.

8.6.1 The classical complement pathway

The main components of the classical complement pathway are proteins called C1–C9. They were numbered in the order in which they were discovered and fortunately, with one exception, they become involved in the chain of events in the same numerical order. Therefore the initiation of the classical complement pathway involves the first component of complement, C1.

(1) Binding of C1 to antibody–antigen complexes initiates the classical pathway

For the complement pathway to be activated antibody must bind to antigens on a solid surface, for example the membrane of a bacterium. When an antibody binds to an antigen, the Fc portion of the antibody alters conformation so that it can bind the first component of complement, C1. This process is also known as **complement fixation**. Not all antibody classes can bind C1: IgM is the most efficient and, in humans, IgG1, IgG2 and IgG3, but not IgG4, can bind C1.

C1 has three components called C1q, C1r and C1s and the complex has the appearance of a bunch of tulips, with six stalks composed of C1q bound to two molecules each of C1r and C1s (see Figure 8.5, and Box 8.3 for more details). Each C1q molecule has a globular protein 'flower head' that is capable of binding to the Fc part of the antibody molecule. At least two of the C1q globular heads must bind to an Fc binding site for C1 to be activated. For IgG molecules, which have only one Fc portion, this means that

Figure 8.5 The classical complement pathway. Two molecules of IgG are needed to bind (fix) C1. Because IgM is a pentamer of Ig, one molecule of IgM can bind C1.

two or more IgG molecules must be bound close together on a particulate surface, such as a bacterium, to activate C1. IgM, which is a pentamer, has five Fc regions and therefore one IgM molecule is capable of activating C1 (① and ② in Figure 8.5). When two or more of the C1q globular heads have bound to Ig Fc regions the C1q undergoes a conformational change, which results in activation of C1r. C1r is an inactive serine protease, which, after activation, cleaves itself to form more active C1r**. C1r** can also cleave C1s, another serine protease, leading to activation of C1s**.

(2) Activated C1 generates products from C4 and C2

C1s** then cleaves another complement component, C4, into a small fragment called C4a and a larger fragment called C4b. The C4b can then attach to the cell membrane or particle to which the Ab/C1 complex is bound. The C4a diffuses away but has important biological properties (see Section 8.6.4). The membrane-bound C4b can now bind the C2 component of complement and when C2 is bound to C4b it in turn is cleaved by C1s** into C2a and C2b. The C2b diffuses away but the C2a remains bound to C4b and the C4b2a forms an active protease complex C4b2a** (③ in Figure 8.5).

C1 consists of 18 polypeptide chains. The three subunits are called C1q, C1r and C1s and there are six C1q, two C1r and two C1s subunits in a C1 macromolecule. Each C1q subunit actually consists of three different polypeptide chains called A, B and C, which twist together in a collagen-like triple helix topped with a globular head.

(3) C4b2a is a C3 convertase

The substrate for C4b2a** is C3, which is a central and key component of all the complement pathways. Because C4b2a** converts inactive C3 into its active form it was originally called **C3 convertase**. The importance of C3 is demonstrated by the fact that individuals who lack C3 suffer from recurrent, life-threatening bacterial infections. C3 binds to C4b2a and is cleaved into C3a and C3b (④ in Figure 8.5). C4b2a** is an enzyme and one molecule of C4b2a** can generate more than 200 molecules of C3b. C3a diffuses away but has important functions (see Section 8.6.4). The C3b generated has two important functions. Most of the C3b binds to the cell to which the antibody is bound and acts as an opsonin, promoting the phagocytosis of the cell. The ability of C3b to act as an opsonin is one of the most important functions of complement. Some of the C3b binds to the C4b2a**, forming C4b2a3b** (④ in Figure 8.5).

(4) C4b2a3b is a C5 convertase

C4b2a3b is a C5 convertase and binds to and cleaves C5 into C5a and C5b (⑤ in Figure 8.5). C5a diffuses away but has important biological activities (see Section 8.6.4) and the C5b binds to the cell surface. The cleavage of C5 is the last enzymatic step in the complement pathway. The later steps are involved in generating pores in the membrane that will result in lysis of the cell. These pores are known as the **membrane attack complex**.

The generation of the C5 convertase is also the last step at which the three pathways differ; the subsequent stages are the same for all the pathways.

(5) C6–9 are also involved in the formation of the membrane attack complex
C5b is quite labile and has about two minutes to bind the next complement component, C6, before it is inactivated (⑥ in Figure 8.6). The C5b6 complex binds C7 and C8 to form a C5b678 complex. During the binding of these molecules some of them undergo conformational changes that result in highly lipophilic structures in C7 and C8 inserting into the membrane. These C5b678 complexes can form small pores (10 Å in diameter) (⑥ in Figure 8.5) in the membrane, which can result in the lysis of some microorganisms and therefore act as a membrane attack complex.

Most cells are not lysed by C5b678 but require the final complement component C9 to be added. Up to 16 C9 molecules can be added to one C5b678 complex and the C9 molecules assemble to form a pore some 100 Å in diameter (⑦ in Figure 8.5). This is usually what is meant by the membrane attack complex (MAC).

(6) Lysis occurs due to osmotic imbalance
The pores formed by the MAC are large enough to allow the passage of ions, small molecules and water but too small for proteins to pass through. There is therefore an influx of water into the cells and a loss of ion balance, which leads to lysis by 'bursting' of the cell.

8.6.2 Lectin pathway

This is very similar to the classical pathway except for the very first steps (Figure 8.6). Instead of antibody binding to an antigen on a pathogen, mannose binding protein (MBP), which is always present in the serum, binds to mannose residues on the surface of pathogens. These mannose residues may be

Figure 8.6 Complement pathways. The classical and lectin pathways result in the generation of the C3 convertase C4b2a and the C5 convertase C4b2a3b. The alternative pathway results in the formation of the C3 convertase C3bBb and the C5 convertase C3bBb3b. Once C5b has bound to the cell surface, the later steps are identical for all three pathways.

components of glycoproteins or polysaccharides. When MBP has bound to the pathogen, a protease called MBL-associated serine protease (MASP) binds to the MBP. The complex of MBP and MASP cleaves C4 into C4a and C4b and C2 into C2a and C2b in the same way that C1s** performs these functions in the classical pathway. Once C4b is generated it binds to the cell surface and the rest of the sequence is the same as for the classical pathway.

8.6.3 Alternative pathway of complement

The important differences between the classical and lectin pathways and the alternative pathway of complement are that C1, C4 and C2 are not involved in the alternative pathway. However, the components C3 and C5–C9 are involved in all pathways. The alternative pathway uses different components to generate C3 and C5 convertases, as described below (Figure 8.6).

(1) Formation of C3 convertase
C3 is quite labile and undergoes a low level of spontaneous hydrolysis to produce C3a and C3b. Most of the C3b generated is itself hydrolysed to an inactive form in the serum within seconds of formation. Some of the C3b binds to the body's own cells where it is inactivated by surface membrane regulatory proteins (see Section 8.6.5). Many microbes lack these regulatory proteins and therefore C3b bound to a microbial membrane is not inactivated. This system provides an elegant, innovative mechanism by which the innate immune system can distinguish self from non-self. Self-tissue is able to inactivate C3b but many microbes are not and go on to be opsonised or lysed by the alternate complement pathway.

If C3b is not inactivated it binds another component of the alternative pathway, factor B. The factor B bound to C3b is then cleaved by another alternative pathway serum protein, factor D, to yield Ba and Bb. Ba diffuses away but Bb remains bound to C3b, forming an active C3 convertase called C3bBb**, which is equivalent to C4b2a of the classical pathway. The C3bBb** is stabilised by another protein called properdin. The C3bBb** can cleave more C3, generating more C3b that can bind to the microbial membrane, which provides a powerful amplification loop. It has been estimated that in this way upwards of a million C3b molecules can be bound to the membrane within five minutes.

This C3b can also perform one of two functions. One of these is to act as an opsonin and just as in the classical pathway the opsonic activity of C3b plays a very important role in eliminating microorganisms. The other role of C3b is as part of the C5 convertase of the alternative complement pathway.

(2) Formation of C5 convertase
Some of the C3b generated by the C3bBb** actually binds to the C3bBb to form C3bBb3b, which is a C5 convertase equivalent to C4b2a3b in the classical pathway. C3bBb3b cleaves C5 into the same C5a and C5b components as C4b2a3b. Once C5a and C5b have been generated the latter stages

of the complement pathway involving formation of the MAC proceed in exactly the same way as for the classical pathway (see above).

8.6.4 Biological activities of complement

The complement cascade, whether it is activated by the classical or alternative pathway, generates a number of molecules that have a variety of biological activities (Figure 8.7). The main activities of the complement components are as follows:

Figure 8.7 Functions of complement components. Complement has many functions in addition to lysis of cells. C3b is an opsonin and when bound to a microbial cell surface can bind to complement receptors (CR1) on phagocytes. Complement components also play a number of roles in inflammation, including causing vasodilation and an increase in vascular permeability (C5a), chemotaxis of neutrophils, monocytes and eosinophils (C5a), and activation of mast cells (C3a, C4a and C5a) and neutrophils (C5a).

- **Cell lysis.** This is mediated by the components C5–9 forming the MAC, as described above.

- **Opsonisation.** C3b is a very important opsonin. Monocytes, macrophages and neutrophils all have receptors for C3b on their surface. These receptors, called complement receptors type 1 (CR1), can bind C3b on the surface of microbes, thereby acting as opsonins and promoting the phagocytosis of the microbe.

- **Chemotaxis.** Many complement components produced by the splitting of inactive precursors have chemotactic activity and therefore play a role in recruiting cells to the site of an inflammatory response. The most active is C5a, which is chemotactic for neutrophils, monocytes and, to a lesser extent, eosinophils. C3a is also chemotactic for eosinophils.

- **Inflammatory mediators.** C3a, C5a and C4a can all act as anaphylatoxins – these are molecules that can activate mast cells and cause their degranulation. C5a can cause release of hydrolytic enzymes from neutrophils. C3a and C5a can cause degranulation of eosinophils.

- **Clearance of immune complexes.** When antibody binds to a soluble antigen, such as a toxin protein secreted by a bacterium, there is the possibility that insoluble antigen–antibody complexes could be formed. These can become trapped in small capillaries and deposited on the capillary walls, leading to a local inflammatory response resulting in immune complex disease (see Chapter 13). Complement has two important roles in preventing this from happening. C3b is able to interact with antigen–antibody and limit the size of the lattice formed by the complex (Figure 8.8). This prevents the complexes from reaching a size where they become insoluble and could be deposited in the blood vessels.

  Erythrocytes and complement can also collaborate in the removal of immune complexes (Figure 8.8). Erythrocytes have CR1 on their surface. Immune complexes with C3b bound to them are able to bind to the CR1 on erythrocytes, which also prevents the immune complexes being deposited in blood vessels. When the erythrocyte with the bound immune complex travels through the liver or spleen, macrophages bind the erythrocyte/immune complex pair through Fc receptors on the macrophages binding to the Ab in the immune complex. The erythrocyte dissociates from the immune complex and returns to the circulation. The immune complex is phagocytosed and destroyed by the macrophage.

8.6.5 Complement regulatory molecules

It is important that a system with the amplification and potentially tissue-damaging properties of the complement system is very tightly regulated. Regulation occurs at two levels. One is to regulate the amount of complement activity and the other is to regulate the site of complement activity. If the site of complement activation is the surface of a pathogen it is important that active complement components do not spread to host cells in the vicinity of the pathogen. Both soluble inhibitors in the plasma or extracellular fluid and membrane-bound inhibitors are important in regulating complement activity.

There are remarkable parallels between regulation of the classical and alternative complement pathways and many regulatory proteins are involved in inhibition of both pathways (Figure 8.9).

Figure 8.8 Immune complex solubilisation and clearance. Immune complexes can result in the fixation of C1 leading, via the classical pathway, to the generation of C3b, which dissociates immune complexes and prevents them getting bigger. The C3b in immune complexes can also contribute to immune complex clearance by binding to CR1 on the surface of erythrocytes. The erythrocytes transport the immune complexes to the liver or spleen where macrophages take up the immune complex and remove it, leaving the erythrocytes to continue circulating in the blood.

Regulation of C3 convertases

Various proteins either inhibit the assembly of C3 convertases or accelerate their dissociation (Figure 8.9):

- **Regulators of C3 convertase assembly.** The classical C3 convertase is C4b2a. Two membrane proteins, complement receptor 1 (CR1) and membrane co-factor protein (MCP), and one soluble protein, C4b binding protein, are able to bind C4b and stop it binding to C2a, thereby inhibiting the formation of C2a4b. Another regulatory protein, factor I, then cleaves C4b, resulting in irreversible inhibition of C4b2a assembly.

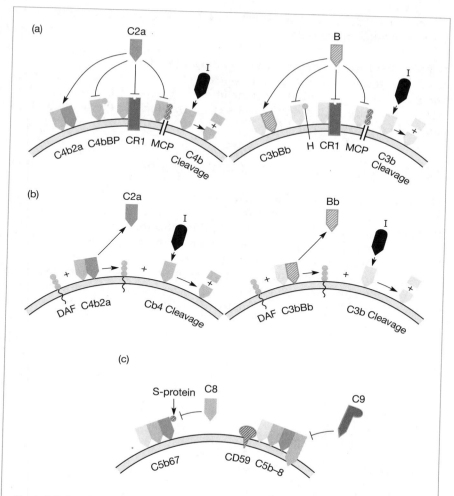

Figure 8.9 Regulation of complement. (a) Prevention of formation of C3 convertase. C4bBP, CR1 and MCP are able to bind to C4b and prevent C2a from binding to C4b in the classical pathway; Factor H, CR1 and MCP bind to C3b and prevent factor B binding to C3b in the alternative pathway, thereby preventing the formation of the C3 convertases C4B2a and C3bBb respectively. C4b and C3b are irreversibly cleaved by factor I. (b) Decay of C3 convertase. Decay accelerating factor (DAF) causes dissociation of C2a from C4b2a or dissociation of Bb from C3bBb and the C4b and C3b are cleaved by factor I. (c) Prevention of MAC (membrane attack complex) formation. S-protein stops the addition of C8 to C5–7 and CD59 stops the addition of C9 to C5–8.

CR1, MCP and a soluble protein, factor H, are able to bind C3b and stop it associating with factor B to form the alternate C3 convertase, C3bBb. Factor I then cleaves C3b, irreversibly inhibiting assembly of C3bBb.

- **Factors causing decay of C3 convertases.** A membrane protein, decay accelerating factor, causes dissociation of C2a from C4b or C3b from Bb, inhibiting the classical and alternate C3 convertases respectively.

Factor I then cleaves the membrane-bound C4b or C3b, making the inactivation irreversible.

Regulation of the membrane attack complex (MAC)
A soluble protein called S-protein (vitronectin) binds the C5b-7 complex and stops it inserting into the cell membrane. A cell surface protein, CD59, inhibits MAC formation on the cell surface by blocking the binding of C8 and C9 to the C5b–7 complex.

8.7 Antibody, complement and the opsonisation of microbes

Although antibody and complement can act independently of each other, the ability of antibody to activate the complement cascade provides the antibody response with a range of additional effector functions with which to combat pathogens, especially extracellular pathogens. In addition to providing extra functions, complement has two other important attributes. It can synergise with antibody in promoting opsonisation, one of the most important functions of these molecules. Phagocytes have receptors for both the Fc portion of antibody (FcR) and C3b (CR1). If a bacterium or other

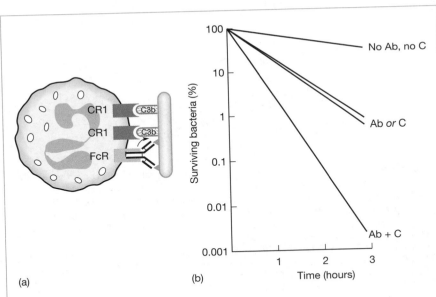

(a) (b)

Figure 8.10 Antibody and complement synergise to opsonise microbes. (a) Antibody bound to a particle (in this case a bacterium) can fix complement, resulting in the attachment of C3b to the surface of the particle. Phagocytes have receptors for both antibody (FcR) and C3b (CR1) and therefore there is very efficient opsonisation and phagocytosis of the antibody/C3b-coated particle. (b) The kinetics of bacterial clearance *in vivo* show the synergistic effects of antibody and complement. It has been estimated that, in the presence of complement, one molecule of IgM may be enough to cause phagocytosis.

organism is coated with both antibody and C3b, the binding and activation of the phagocyte are increased synergistically (Figure 8.10); that is, the combination is greater than can be achieved by either component alone. This results in much more effective clearance of microbes.

Secondly, complement provides a powerful amplification system so that a relatively small amount of antibody–antigen complex can activate many complement molecules. IgM is a pentamer and therefore one molecule of IgM can fix complement. Because of this a single molecule of IgM, in the presence of complement, can cause the opsonisation and phagocytosis of a bacterium. It should be apparent what a powerful combination antibody and complement is.

8.8 Antibody-dependent cell-mediated cytotoxicity (ADCC)

There is one final set of effector mechanisms involving antibody that collectively are called antibody-dependent cell-mediated cytotoxicity, or ADCC for short. Many cell types have FcR on their surface and are therefore able to bind the Fc part of antibody. These cells include neutrophils, macrophage/monocytes, eosinophils and natural killer cells. Therefore if antibody has bound to antigen on a cell surface, effector ADCC cells can bind to the Fc part of the antibody and kill the antigen-bearing cell (Figure 8.11). It is important to realise that killing by ADCC is not the same as phagocytosis. The targets for ADCC are generally too big to be phagocytosed and the killing is extracellular. In phagocytosis the microorganism is ingested and killed intracellularly.

The role of ADCC *in vivo* has been difficult to elucidate and has been presumed from *in vitro* studies. Host cells that have been infected by viruses or intracellular dwelling bacteria or parasites may express antigens derived from the pathogen on their cell surface and therefore be recognised by antibody and be susceptible to ADCC primarily performed by macrophages and NK cells. Additionally eosinophils have been shown to kill helminthic worms coated with antibody *in vitro*. Both IgG and IgE can promote this type of ADCC (Figure 8.11).

8.9 Summary

- Antibody is able to provide protection by both neutralising toxins produced by pathogens and promoting the destruction and elimination of pathogens. Antibody can agglutinate microorganisms promoting their phagocytosis or act as an opsonin, also promoting phagocytosis.

- Phagocytes kill phagocytosed microorganisms through lysosome-independent and lysosome-dependent mechanisms. The former involves the generation of nitric oxide and oxygen radicals and the latter involves the generation of chlorine products and the action of defensins and proteolytic enzymes.

Effector mechanisms: (1) Antibody-mediated responses

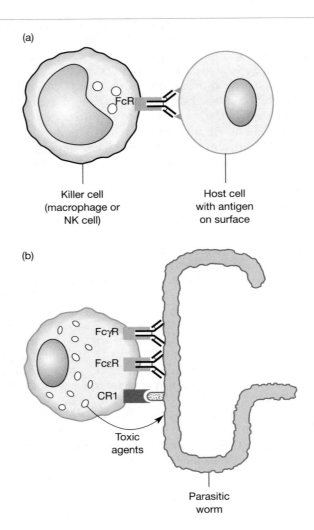

Figure 8.11 Antibody-dependent cell-mediated cytotoxicity (ADCC). (a) Host cells with antigens on their surface may bind antibody. These antigens may be derived from intracellular pathogens or may be tumour-specific antigens (see Section 15.4). Killer cells with FcRs bind the antibody and kill the host cell. The killer cells in this situation are macrophages and natural killer (NK) cells; the killing is extracellular and not by phagocytosis since often the host cell is too big to be phagocytosed. (b) A type of killing of parasitic worms has been demonstrated *in vitro*. Eosinophils have FcRs for IgE and IgG and therefore are able to attach to worms that have antibody specific for parasite antigens on their surface. The eosinophils release toxic agents, especially major basic protein, peroxidase and eosinophil cationic protein, which kill the worm.

- The complement system is a series of serum proteins that have a number of biological functions. There are three pathways of complement activation: the classical pathway, the alternative pathway and the lectin pathway.

- The many products generated as a result of complement activation have a number of different biological activities. Complement can generate a membrane attack complex that punches holes in cells, thereby killing them. Complement can also act as a powerful opsonin, acting independently or in concert with antibody to promote phagocytosis. Many complement components act as inflammatory mediators. Complement also acts to solubilise immune complexes and promote their clearance from the blood.

- Antibody can also promote ADCC in which different cell types kill other cells. These may be host cells infected with bacteria or viruses or helminths, which can be killed by eosinophils via ADCC.

CHAPTER NINE

Effector mechanisms: dealing with pathogens *in vivo*
(2) Cell-mediated immunity

Learning objectives

To know how cytotoxic CD8 T cells are generated and kill infected host cells. To understand the events contributing to a delayed-type hypersensitivity reaction. To appreciate the differing cost to the host of the different types of effector response.

Key topics

- Cytotoxic CD8 T cells
  - ○ Generation of cytotoxic T cells
  - ○ Granule-mediated killing
  - ○ Fas-mediated killing
- Delayed-type hypersensitivity
  - ○ Cellular features
  - ○ Migration of effector Th cells
  - ○ Recruitment of monocytes
  - ○ Activation of monocytes/macrophages
- Consequences to the host of different effector mechanisms
- Two types of helper T cell

9.1 Introduction

Antibody is essential for protection against a wide variety of bacteria, viruses and parasites. There are a number of immunodeficiency diseases in which babies are born with genetic mutations that render them unable to make normal antibody responses, and some are totally unable to make antibody. Before treatment became available, these babies invariably died of overwhelming infection, especially with pyogenic (causing fever) bacteria, such as *Staphylococcus aureus, Haemophilus influenzae* and *Streptococcus pneumoniae*, which have a polysaccharide coat that makes them resistant to phagocytosis in the absence of antibody. These immunodeficiencies illustrate the importance of antibody in fighting infection.

Unfortunately there are some infectious agents where antibody is not effective. These are intracellular pathogens, which can be bacteria, viruses, fungi or protozoa (Table 9.1). Because these pathogens, live intracellularly they are not accessible to antibody, although antibody, can still limit the spread of these organisms if spreading involves an extracellular phase. Antibody can also provide protection against future infection by binding to intracellular pathogens when they first arrive in the body, thereby preventing them from infecting cells (see Section 8.3). Mucosal IgA is particularly effective at preventing viruses from infecting through mucosa. However, upon first infection with these pathogens you do not have antibody against them and by the time antibody is made the pathogens have had time to infect cells and hide in them so that the pathogens are no longer accessible to the antibody. Two other types of effector response are generated to deal with these intracellular pathogens. These are the production of CD8 cytotoxic T cells (Tcs) and the generation of delayed-type hypersensitivity (DTH) responses. These responses do not involve antibody and have been called cell-mediated responses or cell-mediated immunity.

Table 9.1 Intracellular pathogens requiring cell-mediated immune responses for elimination

Pathogen	Disease	Response
Bacteria		
Mycobacterium spp.	Tuberculosis, leprosy	DTH
Legionella pneumophilia	Legionnaires' disease	DTH
Listeria monocytogenes	Meningitis	Tc
Chlamydia psittaci	Psittacosis	Tc
Fungi		
Pneumocystis carinii	Pneumonia	DTH
Cryptococcus neoformans	Meningitis	DTH
Viruses		
Herpes simplex virus	Cold sores	DTH
Measles virus	Measles	Tc
Influenza virus	Influenza	Tc
Protozoa		
Plasmodium berghei	Malaria	DTH, Tc
Trypanosoma	Sleeping sickness	Tc

Tc, CD8 cytotoxic T lymphocyte; DTH, delayed-type hypersensitivity.

9.2 Cytotoxic T cells

Cytotoxic T cells are CD8 T cells that are able to recognise antigenic peptides presented by class I MHC on the surface of a cell. As their name implies, Tcs are able to kill the cells expressing antigen class I MHC. The cytotoxic T cells are sometimes called killer cells and the cell being killed is called the target cell (Figure 9.1). However, CD8 T cells that have not been stimulated by antigen before are not cytotoxic. Just as B cells have to proliferate and differentiate after their first encounter with antigen to become antibody-producing plasma cells, so CD8 T cells also have to proliferate and differentiate after their first encounter with antigenic peptide/class I MHC to become Tcs.

9.2.1 Generation of cytotoxic T cells

Early *in vitro* studies suggested that the generation of cytotoxic T cells was fairly straightforward. The first requirement, as with any other specific immune response, was stimulation of the CD8 T cells by antigen. This required that the CD8 T cell recognised the antigenic peptide/class I MHC for which their TcR was specific. This caused expression of IL-2 receptors and IL-2 alone was able to cause the CD8 T cells to proliferate and differentiate into cytotoxic T cells. Other cytokines, such as IFNγ, IL-4 and IL-10, were not essential but made the process more efficient. It was assumed that *in vivo* the cytokines would be provided by CD4 T cells.

However, when the generation of Tcs in response to viral infections was examined, it was found that Tcs could be generated against some, but not all, viruses in the absence of CD4 T cells. There are therefore at least two pathways for generating Tcs: one is CD4-dependent and the other is CD4-independent. Both of these pathways involve activation of naive CD8 T cells by dendritic cells.

Dendritic cells have two useful features that make them efficient at stimulating CD8 T cells. They express high levels of class I MHC and are

Figure 9.1 CD8 T cell-mediated cytotoxicity. The TcR of the cytotoxic T cell (Tc) recognises antigenic peptide (e.g. from a virus) being presented by a target cell. The Tc is stimulated to deliver death signals to the target cell, which is killed.

therefore able to present antigenic peptides to CD8 T cells. Furthermore dendritic cells can take exogenous antigen and present it on class I MHC; usually exogenous antigen goes to class II MHC – see Section 4.5.2. This means that the dendritic cell does not actually have to be infected with the virus to present antigen to CD8 T cells, which is a very useful feature since not all viruses infect dendritic cells.

Dendritic cells have to be activated to express co-stimulatory molecules, such as CD80 and CD86, to be able to stimulate CD8 T cells to proliferate and differentiate into Tcs, and it is how the dendritic cells are activated that determines whether Tc generation is CD4 T cell-dependent or -independent (Figure 9.2). Some viruses are able to activate dendritic cells directly through stimulation of toll-like receptors on DCs or through the induction of an inflammatory response. These viruses can therefore stimulate Tc generation in a CD4-independent manner. The initial activation of DCs is thought to take place at the site of infection. The activated DCs migrate to the T-cell area of the draining lymph node or spleen, where they can encounter CD8 T cells, and if the TcR on the CD8 T cell is specific for the viral antigen/class I MHC the dendritic cell can activate the CD8 T cell to become a Tc.

Other viruses are not able to directly activate DCs and require CD4 T cells to do so. In addition to expressing class I MHC, dendritic cells also express class II MHC and can present viral antigens to CD4 T cells. CD4 T cells that are specific for the viral antigen/class II MHC are activated to express the cell surface molecule CD154, which binds to its ligand, CD40, on the surface of the DC, resulting in activation of the DC, to express CD80 and CD86, which acts as a co-stimulus for the CD8 T cell and results in proliferation of the CD8 T cell and its differentiation into a Tc. The CD4 T cell itself is stimulated to produce IL-2, which promotes the proliferation of both the CD8 T cells and CD4 T cells, which both express IL-2 receptors after antigen stimulation. The CD4 T cells, will differentiate into Th cells, which secrete other cytokines, such as IFNγ, which can further promote the differentiation of CD8 T cell into Tcs. It can be seen that the ability of the same DC to present antigen to, and stimulate, both CD4 and CD8 T cells provides an efficient process for interaction between cells via contact and cytokines. Even with viruses that can stimulate CD4-independent responses, the normal immune response is thought to involve CD4 T cells.

9.2.2 Mechanisms of CD8 T cell-mediated cytotoxicity

CD8 T cells can kill by two different mechanisms. The first mechanism is via granule exocytosis and the second involves the Fas pathway (Figure 9.3).

Granule exocytosis
CD8 T cells contain granules within their cytoplasm. These granules contain a number of proteins that can cause the lysis of target cells. Two important proteins found in granules are perforin and serine proteases called granzymes.

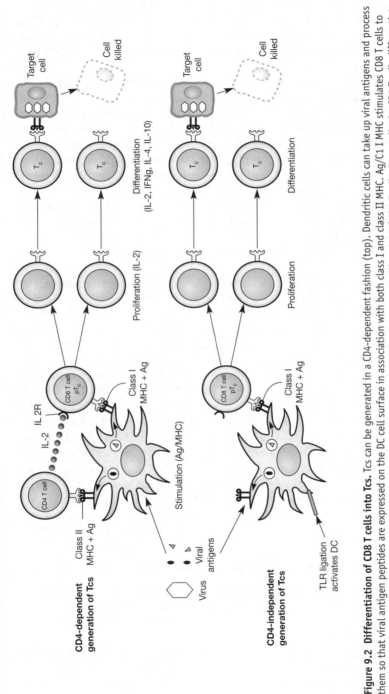

Figure 9.2 Differentiation of CD8 T cells into Tcs. Tcs can be generated in a CD4-dependent fashion (top). Dendritic cells can take up viral antigens and process them so that viral antigen peptides are expressed on the DC cell surface in association with both class I and class II MHC. Ag/C1 I MHC stimulates CD8 T cells to express receptors for IL-2 and Ag C1 II MHC stimulates CD4 T cells to produce IL-2 which causes proliferation of CD8 T cells. Eventually the CD8 T cells differentiate into Tcs, a process that is promoted by IFNγ, IL-4 and IL-10 although these are not essential. Tcs can also be generated in the absence of CD4 T cells (bottom). This requires that the DCs are activated through recognition of pathogen products, e.g. through toll-like receptors (TLRs). The activated DCs express co-stimulatory molecules and can stimulate CD8 T cells by expression of antigen peptides on their class I MHC to differentiate into Tcs.

Figure 9.3 Two pathways of CD8 T cell cytotoxicity. In the Fas pathway, Fas-ligand on the cytotoxic T cell cross-links Fas on the target cell activating the Fas death pathway, leading to apoptosis of the target cell. The granule exocytosis pathway involves degranulation of the Tc and release of granule contents (perforin and granzymes), which activate apoptosis in the target cell.

- **Perforin.** This is a 65 kDa protein with structural homology to the complement component C9. Like C9 it is able to form pores in cell membranes.

- **Granzymes.** These are a series of serine esterases that cleave proteins at serine residues. There are many granzymes in cytotoxic T cell granules, although the specificity of their enzymatic activity appears similar.

Cytotoxicity through granule exocytosis can be divided into four stages:

1. **Recognition and binding of the target cell.** For cytotoxicity to be triggered in the cytotoxic T cell, the cytotoxic T cell must recognise antigen/class I MHC on the target through specific binding of its TcR to antigen/MHC. The requirement that the TcR must bind its specific antigen ensures that only target cells bearing antigen for which the cytotoxic T cell is specific will be killed. The binding of the cytotoxic T cell to the target is promoted by binding of other molecules on the cytotoxic T cell molecules on the target cell. The most important of these other molecular interactions involve CD2 on the T cell binding to CD58 (LFA-3) on the target cell and CD11a/18 (LFA-1) on the T cell binding to CD54 (ICAM-1) on the target cell.

2. **Delivery of the lethal hit.** Following triggering of the cytotoxic T cell, the granules in the T cell move towards the site of attachment to the target cell. The granules fuse with the membrane of the cytotoxic T cell and the contents of the granules are released into the intracellular space between the cytotoxic T cell and the target cell.

3. **Death of the target cell.** Perforin is released as monomers but in the presence of calcium the monomers insert into the target cell membrane

and polymerise to form pores of about 50 Å in diameter. For some target cells this formation of pores is enough to cause osmotic lysis in a manner analogous to that produced by complement. However, many target cells are not killed by perforin alone and killing requires the presence of granzymes.

Granzymes are thought to enter the target cell through the perforin pores. Granzymes cleave proteins in the target cells, which, among other things, results in activation of apoptosis pathways in the target cell. The target cell then undergoes programmed cell death.

4. **Recycling of the cytotoxic T cell.** Once the cytotoxic T cell has delivered the lethal hit it can detach from the target cell and is then capable of killing other target cells bearing the specific antigen/class I MHC. The cytotoxic T cell remains bound to the target cell for 10–15 minutes and the death of the target cell can take place at any time from 5 minutes to 3 hours from the time of detachment of the cytotoxic T cell.

Fas-mediated cytotoxicity

Fas is a death molecule that can be expressed on many cell types. Its expression is also stimulated by a variety of signals. If Fas on a cell interacts with Fas-ligand (Fas-L) the Fas is cross-linked and this activates pathways leading to apoptosis of the cell. Binding of the TcR of CD8 cytotoxic T cells to its specific Ag/MHC class I causes expression of Fas-L. Fas on the target cell will be cross-linked by the Fas-L on the cytotoxic T cell and apoptosis and death of the target cell will be triggered (Figure 9.3). Not all cells express Fas, so this mechanism of cytotoxicity is not effective against all cells.

Cytotoxic T cell responses provide important protection against many viral infections and the magnitude of the cytotoxic T cell response can be huge. It is now possible to measure the number of CD8 T cells that are specific for a particular viral antigen in association with class I MHC. These studies have shown that during some viral infections, up to 20% of all the CD8 T cells in a person's bloodstream are specific for antigens on the virus. This is a truly remarkable response but sometimes even this is not enough to eliminate all intracellular pathogens. In these cases a delayed-type hypersensitivity response may be required.

9.3 Delayed-type hypersensitivity

Delayed-type hypersensitivity responses are often generated against pathogens that live inside macrophages themselves. It might seem that the last place a pathogen would want to live is inside a phagocytic cell, one of whose functions is to phagocytose and kill microbes. However, many pathogens have evolved a variety of mechanisms to avoid being killed by macrophages (see Box 9.1). Not all pathogens that stimulate delayed-type hypersensitivity responses live in macrophages and many live in other cell types (Table 9.2).

The aims of a delayed hypersensitivity response are really quite simple. They are to:

Table 9.2 Cells types occupied by intracellular pathogens stimulating DTH responses

Pathogen	Disease	Cell type occupied
Bacteria		
Mycobacterium tuberculosis	Tuberculosis	Macrophages
Mycobacterium leprae	Leprosy	Endothelial cells, Schwann cells
Legionella pneumophila	Legionnaires' disease	Macrophages
Rickettsia prowazekii	Typhus fever	Endothelial cells
Protozoan parasites		
Leishmania spp.	Leishmaniasis	Macrophages

BOX 9.1: HOW PATHOGENS AVOID BEING KILLED BY PHAGOCYTES

Pathogens have evolved many ways of avoiding being killed by phagocytes. *Listeria monocytogenes*, which can cause meningitis in newborns, has evolved the ability to escape from the phagosome into the cytoplasm of the cell where it can multiply and infect other cells. Other bacteria are able to limit the involvement of lysosomal contents in bacterial killing in a number of ways. *Salmonella typhi* and *Mycobacterium tuberculosis* prevent the phagosome fusing with the lysosome, thereby preventing the bacteria from being exposed to the lysosomal contents. *Mycobacterium* spp. also produce NH_4^+, which neutralises the acid pH of lysosomes, and they can neutralise the proton pump that normally is involved in lowering the pH of the phagolysosome. Some bacteria are resistant to killing by lysosomal contents.

Other pathogens produce factors that interfere with killing by reactive oxygen products. Two enzymes produced by phagocytes are superoxide dismutase (SOD) and catalase. SOD converts superoxide to H_2O_2 in the following reaction:

$$2O_2^- + 2H^+ \xrightarrow{\text{SOD}} H_2O_2 + O_2$$

Catalase then converts hydrogen peroxide to water and oxygen:

$$2H_2O_2 \xrightarrow{\text{Catalase}} 2H_2O + O_2$$

Although normally produced by the phagocyte, many microorganisms produce these enzymes, thereby speeding up the inactivation of O_2 and H_2O_2 and protecting the bacteria.

Another approach to resistance to oxygen radicals is used by *Mycobacterium leprae*, which coats itself with phenolic glycolipid, a molecule that scavenges reactive oxygen radicals, thereby inactivating them.

Finally some organisms, such as *Mycobecterium tuberculosis*, can kill macrophages. This is a mixed blessing since the organism prefers to live and multiply within macrophages. The bacterium is protected from phagocytosis but has lost its home and may be exposed to extracellular defence mechanisms such as antibody.

- recruit monocytes to the site of infection;
- keep monocytes and tissue macrophages at the site of infection;
- activate the monocytes and macrophages to kill the intracellular organisms.

These acticities are all controlled by CD4 T cells, which must be activated to become Th as is any specific immune response (Figure 9.4).

9.3.1 Cellular features of the DTH reaction

Activation of CD4 T cells
Pathogens living in a tissue will shed antigens, which can be picked up by tissue dendritic cells. The dendritic cells pick up exogenous antigen and process it. They are stimulated to leave the tissue site and migrate via the lymphatic vessels to the lymph nodes draining the site of infection. As they migrate, the dendritic cells up-regulate class II MHC and CD80/86 expression on their cell surface and down-regulate their antigen-processing

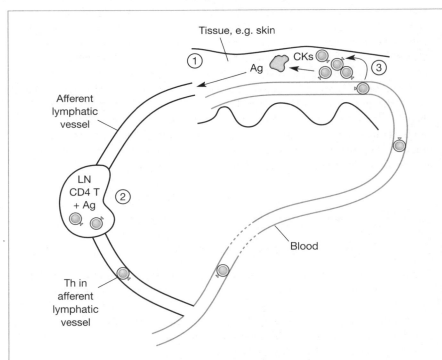

Figure 9.4 CD4 T cells and delayed-type hypersensitivity. ① Antigen in the tissue enters afferent lymphatic vessels and is transported to a lymph node. ② DCs present the antigen to CD4 T cells, which are stimulated to become Th. Th cells leave blood vessels at sites of inflammation and enter the tissue. ③ The Th cells recognise antigen being presented on the surface of monocytes in association with class II MHC, and the Th cells release TNFα, which acts on the endothelium to increase the recruitment of monocytes and Th cells from the blood. The Th cells also secrete IFNγ and IL-2, which, together with TNF, activate the tissue macrophages and recruited monocytes to become more efficient at killing intracellular pathogens.

machinery. The differentiated dendritic cells in the lymph nodes are often referred to as immunostimulatory dendritic cells.The dendritic cells then activate CD4 T cells that are specific for the antigenic peptide being presented by class II MHC on the dendritic cells. The CD4 T cells proliferate and differentiate into effector Th cells, which are sometimes referred to as T_{dth} cells.

Migration of effector Th cells

The Th cells leave the lymph node via the efferent lymphatic vessels, eventually entering the bloodstream via the major lymphatic ducts. It is important that the effector Th cells can return to the site of the infection to carry out their effector functions. This is achieved by the expression of new adhesion molecules on the effector Th cells. There is evidence that the lymphoid tissue in which the Th cells are activated affects which adhesion molecules are induced and which site the effector Th cell will go to (Figure 9.5). If the Th cell is activated in a cutaneous lymph node draining the skin, the Th cell expresses an adhesion molecule called cutaneous lymphocyte-activation antigen (CLA). Th cells that are activated in a mucosal lymph node express different adhesion molecules called LPAM-1 and L-selectin.

The Th cells are able to migrate specifically to cutaneous or mucosal sites because of expression of adhesion molecules on the endothelium due to cytokines being made at the site of the DTH reaction. TNFα and IL-1 made by macrophages stimulate local endothelial cells to express adhesion molecules. E-selectin is induced on cutaneous endothelium and is a specific ligand for CLA on the Th cell, thereby enabling the Th cell to bind to the endothelium and enter the DTH site. Similarly MadCAM-1 is induced on mucosal endothelium at sites of inflammation. This binds both LPAM-1 and L-selectin expressed on mucosally stimulated Th cells, enabling them to home back to the mucosal site of DTH.

The selective up-regulation of different adhesion molecules in cutaneous and mucosal sites that enables the effector Th cell to return to the site of antigen is important because many pathogens infect either mucosal or cutaneous sites but not both. Therefore Th cells specific for a mucosal pathogen antigen are required in the mucosa and not the skin and vice versa for skin pathogen-specific Th cells.

Recruitment and retention of monocytes

The expression of new adhesion molecules on endothelium at the site of DTH reactions also promotes the recruitment of monocytes from the bloodstream. Important among these adhesion molecules is vascular cell-adhesion molecule-1 (VCAM-1), which is expressed on endothelial cells and binds to the CR3 on monocytes.

Monocytes that are recruited to the site of a DTH reaction are subsequently stimulated by cytokines to differentiate into macrophages, which are then activated. These macrophages are retained at the site of the DTH by a cytokine called macrophage inhibition factor, which is secreted by the effector Th cells.

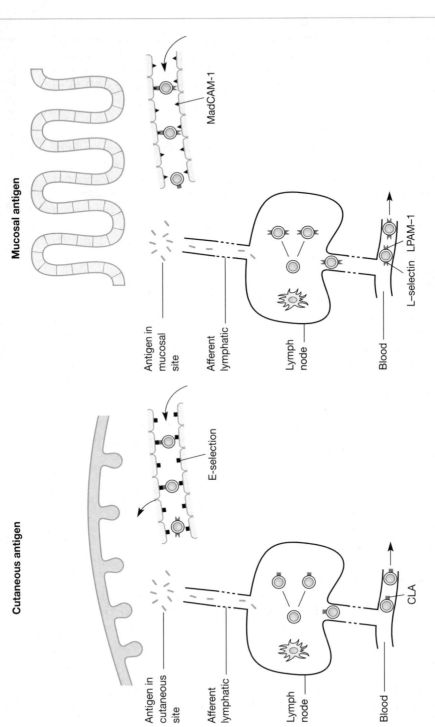

Figure 9.5 Migration of Th. Cutaneous antigen goes to draining cutaneous lymph nodes where CD4 T cells are stimulated. CD4 T cells stimulated in cutaneous lymph nodes express an adhesion molecule called cutaneous lymphocyte-activation antigen (CLA). This binds to E-selectin, which is expressed on cutaneous endothelial cells in response to inflammatory mediators, thereby guiding the Th back to the cutaneous site. Mucosal antigen goes to mucosal lymph nodes and CD4 T cells stimulated here express the adhesion molecules L-selectin and LPAM-1, which bind to MadCAM-1 expressed on mucosal endothelium and guide the Th to the mucosal site.

Activation of macrophages

The macrophages and Th cells that are recruited to a DTH site are able to stimulate each other and provide an amplification loop that increases the magnitude of the DTH response (Figure 9.6). This occurs as follows: The Th cells secrete cytokines that activate the macrophages. The most important of these cytokines is IFN-γ but TNFα and IL-2 can contribute to the macrophage activation in the presence of IFN-γ. The activated macrophages up-regulate class II MHC expression on their cell surface. They are also good antigen-processing cells and can therefore present antigen on their class II MHC to the Th cells. The Th cells are stimulated by recognising antigen/class II MHC on the macrophages to secrete more cytokines that further activate the macrophages, and so the response is amplified. The requirement for the continued stimulation of Th cells to maintain cytokine production also provides a level of control of the response. Once the antigen has been eliminated the macrophages can no longer stimulate the Th cells to produce the cytokines necessary to maintain the response and the response will gradually disappear.

Elimination of pathogens in DTH reactions

Although activated macrophages are better at stimulating Th cells, the main reason for activating the macrophages is to increase their ability to kill phagocytosed pathogens. Activated macrophages have increased levels of nitric oxide and oxygen radicals and higher secretion of proteolytic enzymes. Pathogens that can avoid killing by less active macrophages become susceptible to the enhanced microbicidal activity of the macrophages and are killed.

9.4 Different effector responses have different costs to the host

The different types of effector responses that the immune system is able to generate are usually effective at protecting us from most pathogens that we encounter. However, the different types of responses can result in more or less damage to the host's own cells and tissues. Antibody responses are generally the least damaging to the host. This is because antibody is targeted directly at the pathogen or its products. Neutralisation of toxins by antibody carries essentially no risk of damage to the host. Similarly, complement-mediated lysis of a pathogen or phagocytosis and killing of a pathogen by a phagocyte will usually not cause damage to the host.

Cytotoxic T cell responses are directed at infected host cells and therefore must involve a degree of damage to the host. Cytotoxic T cells seem to provide the best protection against chronic non-lytic viral infections where killing of virally infected host cells limits the replication of the virus and on balance can be beneficial. Acute lytic viruses are better dealt with by antibody because they replicate rapidly and kill the host cell. There is not much point in a cytotoxic T cell killing the host cell if the virus will do so shortly

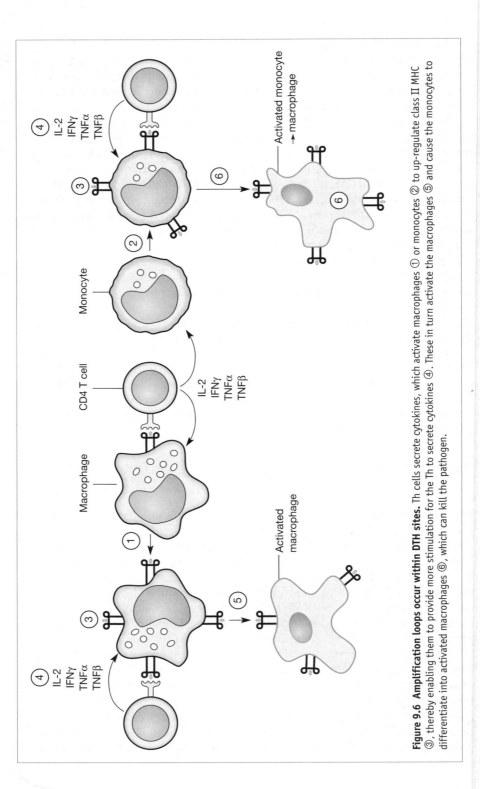

Figure 9.6 Amplification loops occur within DTH sites. Th cells secrete cytokines, which activate macrophages ① or monocytes ② to up-regulate class II MHC ③, thereby enabling them to provide more stimulation for the Th to secrete cytokines ④. These in turn activate the macrophages ⑤ and cause the monocytes to differentiate into activated macrophages ⑥, which can kill the pathogen.

anyway. In some cases the cost of killing virally infected host cells may be too high. This is presumably why neuronal cells do not express class I MHC. Neuronal cells cannot be replaced and on balance less damage may be caused to the host by allowing a non-lytic virus to replicate in neurons rather than killing them.

DTH responses carry the greatest potential risk of damage to the host. Although in many cases the DTH response eliminates the pathogen with little damage, if the response is too great the reactive chemicals and proteo-lytic enzymes produced by the macrophages are secreted and can damage host cells and tissue. In severe cases DTH responses do not eliminate the infection and the DTH response becomes chronic, leading to the formation of **granulomas**. Granulomatous lesions involve chronic stimulation of macrophages with the deposition of fibrin and formation of giant epitheloid cells (Figure 9.7). The excessive production of degradative enzymes by macrophages leads to extensive tissue damage and fibrin can 'wall off the area', leading to a lack of oxygen and tissue necrosis. However, the fibrin walls may limit the spread of microbes from the site.

The importance of the different types of immune response to different pathogens is illustrated by the types of infection that are common in people with genetic (congenital) deficiencies of a particular component of the immune system (see Box 9.2).

9.5 Two major types of CD4 helper T cell

The finding that Th produce a large variety of cytokines that are involved in different types of effector responses led to speculation that different types of Th might exist. Two types of Th have been identified, called Th1 and Th2, and it is likely that there are other types. Th1 and Th2 cells secrete different patterns of cytokines and therefore have different regulatory functions. The cytokines secreted by Th1 and Th2 cells are listed in Table 9.3. Th1 are involved in the generation of Tcs and DTH reactions and promote some classes of antibody. Th2 cells promote other classes of antibody and regu-late responses especially involving mast cells and eosinophils.

9.5.1 Factors controlling Th differentiation

The way in which a naive CD4 T cell differentiates into a Th1 or Th2 effector cell is crucial to producing the right sort of response to a pathogen. In real life, most pathogens probably stimulate a mixture of responses and there are elements of both Th1 and Th2 responses. However, in many cases one partic-ular type of response will predominate and if the response persists for any length of time it may polarise towards a pure Th1 or Th2 response.

There are many factors that influence the differentiation of CD4 T cells. These include dose and timing of antigen, type of antigen-presenting cell and co-stimulatory molecules (see Box 9.3 for more details). However, the

Figure 9.7 Granuloma formation. If the macrophages are unable to eliminate the pathogen in a DTH reaction there will be continuing stimulation of Th cells, which in turn continue to stimulate the macrophages ①. Chronically stimulated macrophages differentiate into epithelioid cells ②, which can fuse to form giant multinucleated cells ③. The mixture of Th cells, macrophages, epithelioid cells and giant multinucleated cells forms, a granuloma ④. Lytic agents and enzymes released by the macrophage-derived cells within a granuloma cause extensive tissue damage and necrosis. In some cases the edge of the granuloma is walled off by epithelioid cells and fibrin, leading to further necrosis due to lack of oxygen.

BOX 9.2: WHAT CONGENITAL IMMUNODEFICIENCY DISEASES TELL US ABOUT THE IMMUNE SYSTEM

Congenital immunodeficiency diseases are caused by mutations in a single gene. These mutations are either inherited or arise spontaneously. These mutations can affect:

- the production of a cell type;
- the production of a molecule;
- the function of a molecule.

Immunodeficiency diseases have been divided into those affecting the specific immune system and those affecting the innate immune system. Diseases affecting the specific immune system are those where the **primary defect** is in the functioning of a B or T cell, and those affecting the innate system are where the primary defect is in a component (cell or molecule) of the innate immune system. However, because of the close interaction between the innate and specific immune systems, diseases that are due to aberrant functioning of one cell type can affect the functioning of other

Examples of congenital immunodeficiency diseases

Primary cell or system affected	Types of infection suffered
Diseases affecting antibody production, e.g. Bruton's agammaglobulinaemia, common variable hypogammaglobulinaemia	Respiratory and GI tract, especially with extracellular pyogenic bacteria such as staphylococci, streptococci, haemophilus
Diseases affecting T cells, e.g. severe combined immunodeficiency (SCID)	Systemic viral infections, especially affecting respiratory and GI tracts: measles, herpes, cytomegalovirus Fungal infections: *Pneumocytis carinii*, *Candida albicans*
Diseases affecting phagocytes: chronic granulomatous disease, Wiscott–Aldridge syndrome	Bacterial infections (especially catalase-positive) affecting GI tract, skin, urinary tract: staphylococcus, *E. coli*. Fungi: *Candida albicans*
Diseases affecting complement: C3 deficiency C5, 6, 7, 8 or 9 deficiency	Pyogenic bacteria such as staphylococci Neisserial infections: gonorrhea or meningitis

cell types. For example there is a group of immunodeficiency diseases called **common variable hypogammaglobulinaemia**. People with this syndrome show defects in the production of one or more antibody classes. In many individuals with this disease, the primary defect is in the CD4 T cells, but, because of the need for Th cells to make antibody, there is a knock-on effect on antibody production.

Despite some of the complexities in identifying the cell or molecule being affected, immunodeficiency diseases have given us a lot of information on the importance of different components of the immune system in protection against different types of pathogens, particularly at the effector stage. Some examples are given in the table.

These diseases helped to establish the importance of antibody and complement in extracellular infections and T cells in intracellular infections. Phagocytes are especially important in extracellular bacterial infections but also operate against some intracellular infections through delayed-type hypersensitivity reactions.

most important factors in determining whether CD4 T cells develop into Th1 or Th2 cells are the cytokines that the CD4 T cells are exposed to.

The role of cytokines in CD4 T cell differentiation

Different cytokines can influence Th differentiation to a greater or lesser extent and it is the mixture of cytokines that will determine which type of Th develops. The cytokines influencing Th differentiation are depicted in Figure 9.8. Many of these cytokines can be produced by cells of the innate immune system; for instance, macrophages and dendritic cells produce IL-12 and mast cells can produce IL-4. This further demonstrates how the innate and specific components of the immune system interact both during the development of specific immune responses and at the effector stage of immune responses (see Chapter 8).

Table 9.3 Cytokines secreted by Th1 and Th2 cells

	Th1	*Th2*
Cytokines selectively secreted by Th1 or Th2	IL-2 IFNγ TNFβ	IL-4 IL-5 IL-6 IL-9 IL-10 IL-13
Cytokines secreted by both Th1 and Th2 cells	IL-3 GM-CSF TNFα	

Although cytokines exert the greatest influence in determining Th1/Th2 differentiation, other factors have been shown to have some influence, although they may well do so by affecting cytokine production.

Antigen dose

In experimental situations antigen dose has been shown to affect Th development, with low doses of antigen favouring Th2 and higher doses Th1 development. Related to this, the frequency of antigen exposure can also have an effect; continuous exposure to antigen favours Th2 development and intermittent exposure biases the response towards Th1.

Antigen-presenting cell

The nature of the antigen-presenting cell may also be influential. Although dendritic cells are the most important cells for activating resting CD4 T cells and therefore initiating immune responses, as the response develops B cells and/or macrophages can become activated and express high levels of class II MHC and CD80 and CD86, therefore enabling them to provide co-stimulatory signals required for activation of resting CD4 T cells. Some evidence indicates that if macrophages assume the major APC role in an immune response this favours a Th1 response, while B cells in the main APC role biases the response towards Th2.

Co-stimulatory molecules

Finally the nature of the co-stimulatory molecules that engage the CD4 T cell may have an effect, with CD80 biasing the responses towards Th1 and CD86 towards Th2.

It should be pointed out that the cytokine influence is dominant in determining Th differentiation. Although the other factors mentioned above can have an influence, the presence of sufficient IL-12 or IL-4 will prevail over other factors, resulting in Th1 or Th2 cells respectively.

The most important cytokine determining Th1 development is called interleukin-12 (IL-12). IL-12 can bind directly to IL-12 receptors on CD4 T cells and signal them to differentiate down the Th1 pathway. It can also stimulate the production of IFNγ by T cells or natural killer cells and the IFNγ promotes Th1 differentiation. In humans, but not in mice, IFNα and IFNβ also promote the differentiation of Th1 cells. IFNα and IFNβ are produced by a variety of cells, including macrophages and fibroblasts, following viral infection. Through the production of IFNα and IFNβ many different cell types that are not part of the specific immune system can promote the production of Th1 cells, which are the Th type required in anti-viral responses.

Figure 9.8 Differentiation of CD4 T cells into Th1 or Th2 cells. The major cytokines controlling differentiation into Th1 cells are IL-12 and IFNγ. IL-12 is produced by DCs and macrophages in response to microbial products such as lipopolysaccharide. IL-12 can act directly on CD4 T cells or stimulate NK cells to produce high levels of IFNγ that act on the CD4 T cells. IL-4 is the major cytokine promoting Th2 differentiation, but its source early on is not clear. One possible source is tissue mast cells.

It is important to appreciate that it is the balance of cytokines that will determine the type of Th generated. This is illustrated by the cytokine interleukin-18 (IL-18) which on its own promotes Th2 responses but in the presence of IL-12 promotes Th1 responses.

9.6 Summary

- The generation of CD8 cytotoxic T cells and delayed-type hypersensitivity responses are two types of specific immune response that involve T cells but not antibody.

- CD8 cytotoxic T cells kill target cells bearing endogenously derived antigen on their class I MHC molecules.

- CD8 cytotoxic T cells kill by two mechanisms: the granule exocytosis and Fas pathways.

- Granule exocytosis involves the release of granule contents of the cytotoxic T cell. The granule contents include perforin and granzymes, which rapidly induce apoptosis in the target cell.

- If target cells express Fas, Fas-ligand expressed by the cytotoxic T cell can activate Fas death pathways in the target cell, also resulting in apoptosis.

- Delayed-type hypersensitivity (DTH) reactions occur against intracellular pathogens where antibody and CD8 cytotoxic T cells may not be fully effective.

- In DTH reactions, antigens from pathogens residing in a tissue site activate CD4 T cells in the draining lymph node to become Th effector cells. The Th cells travel through the blood and enter the tissue where the pathogen is located.

- At the site of infection Th cells release cytokines, which cause the recruitment of monocytes to the area from the blood and then activate the monocytes and any local tissue macrophages.

- The activated monocytes and macrophages have increased ability to kill phagocytosed pathogens and may kill them with little damage to host tissue.

- Sometimes DTH reactions can cause extensive tissue damage in severe acute reactions or in situations where the reaction becomes chronic, leading to granuloma formation.

- CD4 T cells can differentiate into at least two functionally different types of helper cell called Th1 and Th2, which secrete different combinations of cytokines.

- Th1 cells promote the generation of Tcs and DTH reactions and some classes of antibody while Th2 cells promote the production of other antibody classes and recruit eosinophil and mast cells to sites of infection.

CHAPTER TEN

Immunological memory and vaccination

Learning objectives

To understand the basis of immunological memory and how this protects against repeated infection with the same pathogen. To know about current and developing approaches to vaccination.

Key topics

- Immunological memory
  - ○ Memory for antibody production
  - ○ T cell memory
- Vaccines
  - ○ Properties of an ideal vaccine
  - ○ Current vaccines
  - ○ Future vaccines

10.1 Immunological memory

One of the features of specific immune responses is known as immunological memory, but what exactly is meant by immunological memory? Immunological memory can best be illustrated by examining antibody responses after a first and second injection with antigen (Figure 10.1). IgM is made typically within a week after the first immunisation. IgG production follows after IgM. If you look at antibody production following a second exposure to the same antigen the IgM response is essentially the same. However, the IgG response shows dramatic differences from the primary injection. The IgG response is **faster**, **bigger** and, although not depicted in Figure 10.1, **better**. The response to the first immunisation with antigen is also called the **primary** response and that to the second immunisation the **secondary** response. Another feature of immunological memory is that it is specific for the particular antigen. In the example in Figure 10.1 you only get a bigger response to a second immunisation with the same antigen; if the second immunisation was with a different antigen to the first you would get a primary response to the second antigen. Immunological memory is a feature of all specific immune responses and therefore involves B cells, CD4 T cells and CD8 T cells.

10.1.1 B cell memory

B cell memory is the best understood feature of immunological memory. Memory B cells are generated only in germinal centres and are B cells that have undergone proliferation, class switch and affinity maturation (Figure 10.1). Because antigen-specific B cells have undergone proliferation before becoming memory B cells, there will be more antigen-specific memory B cells than there were B cells before the encounter with antigen. This will contribute to a bigger and faster antibody response. Memory B cells have already undergone antibody class switch, and therefore have IgG, IgA or IgE on their surface, which enables them to be distinguished from naive B cells that have not seen antigen before and that express IgM plus IgD on their surface. It also means that memory B cells do not have to undergo class switch when restimulated by antigen, which speeds up the secondary response. The fact that memory B cells have undergone affinity maturation means that the secondary response will be better, i.e. higher-affinity antibody production, than the primary response.

Memory B cells still have to proliferate and differentiate into plasma cells before they secrete antibody and this process requires help from CD4 T cells. Memory T cells circulate through the lymphatic and blood system as described in Chapter 6 and the way in which they interact with CD4 T cells in lymphoid tissue is thought to be the same as described in Section 7.3. However, there are some differences in stimulation of memory B cells, one of them being that soluble antigen is good at stimulating memory B cells while antigen needs to be presented on the surface of a macrophage to stimulate naive B cells.

Not only is the differentiation of memory B cells into plasma cells dependent of CD4 T cells but also the actual development of memory B cells is dependent on CD4 T cells.

10.1.2 CD4 T cell memory

This is the least well understood facet of immunological memory, a somewhat paradoxical situation given the central role of CD4 T cells in regulating specific immune responses. One of the main problems is that there are no definitive ways of identifying quiescent memory CD4 T cells in the same way that surface IgG, IgA or IgE can be used to identify memory

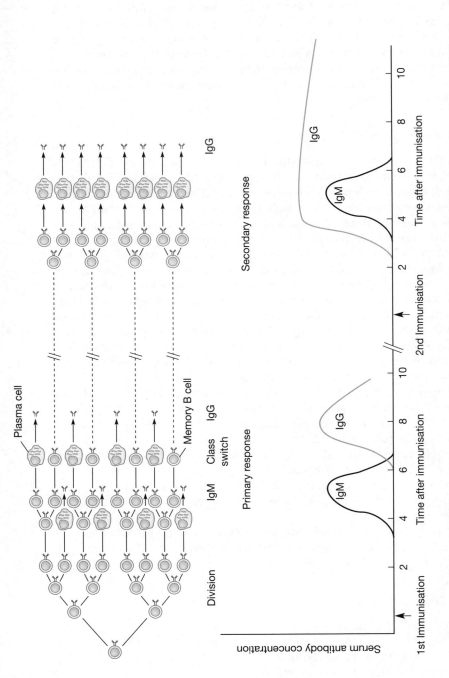

Figure 10.1 Upon the first encounter with an antigen there are relatively few B cells with surface Ig specific for the antigen. The antigen-specific B cells are stimulated to proliferate. After a few rounds of cell division some of the B cells differentiate into plasma cells and secrete IgM. Other B cells continue proliferating and then undergo class switch to IgG. Some of these B cells differentiate into plasma cells and secrete IgG, while other B cells become memory cells. If the same antigen is encountered again there are more memory B cells specific for the antigen and these quickly proliferate and differentiate into plasma cells, giving a

B cells. One of the questions for which there is evidence both for and against is whether memory CD4 T cells are **functionally** different from naive CD4 T cells or whether CD4 T cell memory is simply quantitative, i.e. upon first encounter with antigen, antigen-specific CD4 T cells proliferate so that when you encounter the antigen again you have more CD4 T cells specific for that antigen so that they can generate more Th faster than in a primary response. There is evidence that memory CD4 T cells have less requirement for co-stimulation to be activated than do naive CD4 T cells, suggesting that memory CD4 T cells are functionally different.

10.1.3 Memory CD8 T cells

During a primary Tc response, the proliferation of CD8 T cells generates large numbers of antigen-specific CD8 T cells, about 90% of which die after the infection is cleared. The remaining 10% of cells provide CD8 memory T cells. These are able to respond more quickly to subsequent infection by providing more cells that can differentiate more quickly into Tcs than naive CD8 T cells. Memory CD8 T cells also produce more cytokines, especially IFNγ, after antigen stimulation than do naive CD8 T cells. Although the generation of Tcs in a primary response is not necessarily dependent on CD4 T cells, the generation of memory CD8 T cells does not occur in the absence of CD4 T cells.

10.2 Vaccines

The understanding that the immune system could protect against further infection with a pathogen provides the basis for vaccination. It has been known since ancient times that there are many diseases that people usually get only once in their lives. These include diseases such as smallpox, measles and mumps. Although nothing was known about the mechanisms (indeed, up until the past few centuries, nothing was known about infectious pathogens) this state of protection against further infection was known as **immunity**. It was also known that immunity was specific – if you got small-pox and survived you would be protected against getting smallpox again but not against getting other diseases. Vaccination was the process of trying to induce immunity without going through the disease process, and some forms of vaccination have been practised since ancient times (see Box 10.1). It is now appreciated that immunity is brought about by immunological memory.

10.2.1 Properties of an ideal vaccine

Vaccination is a way of manipulating the immune system to provide protection from disease caused by a pathogen without subjecting the person or animal to the disease. To be effective, a vaccine should stimulate the right type of immune response in the right anatomical site. Depending on the pathogen, antibody, CD8 T cells, delayed-type hypersensitivity (DTH) or a combination of these would be appropriate. For many pathogens the induction of secretory IgA at the appropriate mucosal site may protect from infection. There are also other properties that an ideal vaccine should have:

- **Safety.** There should be no side-effects from the vaccine and no risk of procedural errors in vaccine manufacture exposing individuals to pathogens or their toxic products.

- **Price.** Cheapness is obviously desirable and almost essential in many parts of the world with high rates of endemic disease and few economic resources.

- **Stability.** Ideally a vaccine should be able to be stored in high ambient temperatures so that it is available in hot climates with limited refrigeration facilities.

- **Ease of administration.** Children, the main target of vaccination, do not like having needles stuck in them and therefore oral vaccines are the most suitable form for administration and also can cut out the cost of materials for injection.

Vaccines have become safer. Variolation, used to induce protection against smallpox (see Box 10.1), carried a 1% death rate, which would now be considered unacceptable. The first true vaccine, using cowpox to vaccinate against smallpox (see Box 10.1), was much safer. Many of the vaccines that have been developed and are still in use today are quite safe and effective, despite being developed when relatively little was understood about the immune system. Vaccines are now available for a wide variety of pathogens (see Table 10.1).

However, there are still problems. Some vaccines carry a small risk of side-effects. Although the actual risk is small there can be a knock-on effect, which is the loss in a population of a phenomenon called **herd immunity**. Herd immunity occurs when the proportion of people in a population that is immune to a pathogen is so high that the pathogen cannot find enough susceptible hosts to infect and cannot survive in the population. Therefore even the minority of individuals who are not immune to the pathogen do not get infected. However, if people stop being vaccinated, or having their children vaccinated, the proportion of susceptible people increases to a level where the pathogen can become re-established in the population and herd immunity is lost.

A second issue with vaccines is that there are still many diseases for which no effective vaccine exists. These diseases, including malaria, tuberculosis and parasitic worms, kill in excess of ten million people annually and some estimates suggest 30% of infant mortality could be prevented by successful vaccination. There is still no vaccine for AIDS.

10.2.2 Vaccines – the past and the present

The situation with cowpox and smallpox is unusual. Other pathogens do not have naturally occurring harmless equivalents in other species that

Many ancient peoples practised a form of vaccination even if they did not know about the existence of infectious microorganisms. They were aware that certain diseases were contagious; that is, they could be transmitted between people and/or animals, and that people who survived their first bout with the disease were protected from getting the disease again. The disease that has probably received the most attention over the millennia is smallpox. This is probably due to the magnitude of the threat posed by the disease and because even the ancient approaches to protection were partially effective. Until its eradication in 1979, as a result of a global vaccination programme organised through the World Health Organization, smallpox had been a major scourge of humankind. It is hard to imagine today the scale of the problem but in medieval times up to 60% of the population were infected with smallpox, of whom some 15% died. This means that about 10% of the population died of smallpox. Those who survived, especially adults, were often severely disfigured and/or blinded. Ancient Iranians, Chinese and Indians practised a technique known as variolation, which involved deliberately introducing material from a smallpox lesion into a scratch on the individual being treated, who was usually a child. Even though the process induced the actual disease, it was known that smallpox caused a less severe illness in children and those that survived variolation would be protected as adults. A variation among the dwellers in Baluchistan was to encourage children with wounds to touch the cowpox skin lesions of infected cows. Camelpox was claimed to be as effective. The idea that cowpox could provide protection against smallpox thus preceded the vaccination strategy of Jenner by centuries.

Another disease that attracted considerable attention was rabies. Romans in the fifth century advocated protecting cattle from rabies by making infected cows swallow the boiled liver of a rabid dog. Valli in the 17th century claimed that incubating saliva obtained from a rabid dog (obtaining this seems a somewhat risky process) with the gastric juices of a frog provided material that protected against rabies. Both of these treatments could be considered types of attenuation in which the virus or toxin was rendered harmless but provided immunity.

Other 'vaccination' procedures may have also inadvertently resulted in attenuation of pathogenic organisms or toxins. Inhabitants of ancient Iran treated (vaccinated?) goats against pneumonia by taking the lungs from infected animals and grinding them with garlic and vinegar. A needle and thread was then incubated in the material before being passed through the ear of the goat. A variation on this, practised in parts of Africa, was to ferment pieces of infected lung in a mixture of bran, millet and a plant extract used for tanning skin, to produce a potion for protection against bovine pneumonia. It was observed that although the treatment sometimes caused fatal pneumonia, where it was effective protection was lifelong.

Edward Jenner

Edward Jenner is credited as being the pioneer of modern vaccination. Variolation had been introduced to England in the early 1700s by Lady Elizabeth Montague who had observed the practice in Turkey. Edward Jenner was a country doctor with a practice in Gloucestershire. He had become aware of the local folklore that milkmaids had clear skin, which was not blemished by the ravages of smallpox. He developed the theory that exposure to cowpox protected the milkmaids from smallpox, although the existence of infectious microorganisms was then unknown. To test his theory he performed in 1798 what would today be regarded as an extremely unethical experiment. He took material from a cowpox lesion and introduced it into a scratch on the arm of a young boy. No illness followed. To test whether the exposure to cowpox material protected against smallpox, Jenner deliberately infected the boy with material from a smallpox lesion. Fortunately the experiment was successful and the boy was completely protected. Although it seems highly unethical in today's context to expose someone deliberately to a potentially deadly infection, it should be remembered that variolation, which involved deliberately infecting children with smallpox, was still being practised, and was considered by many to be worth the risk. The term 'vaccination' comes from vaccinia, the cowpox virus.

Table 10.1 Some of the current available vaccines

Organism	*Type of vaccine*
Viruses	
Measles	Attenuated virus
Mumps	Attenuated virus
Rubella (German measles)	Attenuated virus
Polio	Attenuated, killed virus
Chickenpox	Attenuated virus
Rotavirus (infant diarrhoea)	Recombinant vaccine
Influenza	Inactivated virus
Hepatitis (A and B)	Recombinant vaccine
Bacteria	
Diphtheria	Toxoid
Pertussis (whooping cough)	Killed bacteria, subunit vaccine
Meningitis – *Neissera meningitidis*	Capsular polysaccharide
– *Haemophilus influenzae*	Polysaccharide-conjugate
Tetanus	Toxoid
Streptococcus pneumoniae	Capsular polysaccharide
Cholera	Killed bacteria

provide a ready-made source of vaccine. Therefore other approaches were needed to develop vaccines, many of which are still in use today:

- **Killed or inactivated pathogen.** This approach, used primarily for viruses, involves taking the virus and treating it with heat or chemicals so that it is no longer infectious. These vaccines stimulate good antibody responses but, because the virus is no longer infectious, do not stimulate cell-mediated immunity. This approach has been used for influenza, rabies, polio (Salk vaccine) and some bacteria including cholera. One problem is the possibility of some pathogen particles surviving the killing or inactivation process; this happened with the polio vaccine in the 1950s.

- **Attenuated pathogen.** Attenuated pathogens are still viable and cause infection but do not cause disease. Attenuation is usually achieved by growing the organism in cells of another species so that the pathogen becomes adapted to cells of the other species and grows poorly in human cells. Because they can infect cells they stimulate cell-mediated immunity as well as antibody. These vaccines have been developed for polio (Sabin vaccine), measles, mumps and tuberculosis, among others. A problem is the possibility of reversion to full pathogenicity, either in the vaccine stock or in the vaccinated individual.

- **Subunit vaccines.** In some situations an antibody response against a particular component of the pathogen is sufficient to provide immunity. Many bacteria produce a polysaccharide coat that prevents phagocytosis in the absence of antibody. Vaccination with the polysaccharide induces antibody, which is enough to provide immunity. This approach has been used against *Haemophilus influenzae*, which causes 'flu, and *Neisseria meningitides*, a cause of potentially fatal meningitis. One problem with these vaccines is that the polysaccharide antigen does not stimulate Th cells and therefore only IgM is produced. To overcome this, the polysaccharide can be conjugated to a protein such as tetanus toxoid. The tetanus toxoid stimulates Th cells, which can help the B cells specific for the polysaccharide to switch to other antibody classes and make a bigger antibody response (Figure 10.2).

  Subunit vaccines can also be proteins as in the case of the vaccine for hepatitis B, where immunisation with the major surface antigen of the virus, called the HbsAg, induces the production of protective antibodies.

- **Toxoids.** Where pathogens cause disease almost solely through the production of toxins it is possible to vaccinate just against the toxin. This prevents disease upon infection and the immune system can then mount a response against the pathogen and eliminate it. To prevent the toxic effects of the toxin upon vaccination, the toxin is treated chemically so that it loses toxicity but retains antigenicity. Examples of toxoid vaccines are those for tetanus and diphtheria.

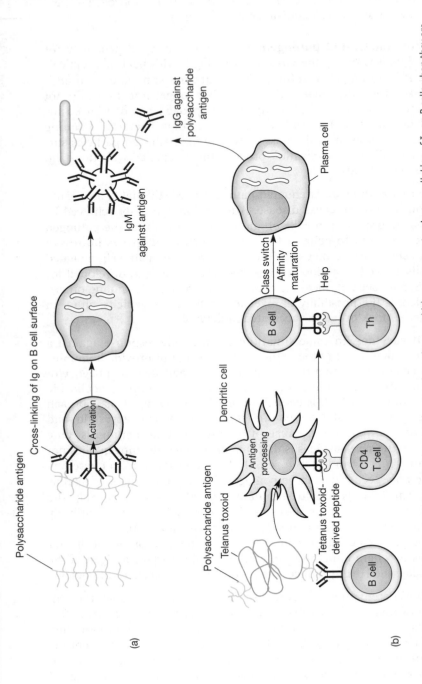

Figure 10.2 Subunit vaccines. (a) Bacterial polysaccharides contain repeating epitopes, which can cause extensive cross-linking of Ig on B cells when they are bound by the Ig specific for the epitope. This is a strong enough stimulus to activate B cells to become antibody-secreting plasma cells in the absence of T cell help. However, because CD4 T cells are not stimulated there is no class switch or affinity maturation and only low-affinity IgM is produced. Because IgM is very efficient at fixing complement this can provide protection against infection by the bacteria. (b) The polysaccharide can be conjugated to a protein such as diphtheria toxoid, to which most people are immunised. The toxoid has T cell epitopes that stimulate toxoid-specific Th, which can help the polysaccharide-specific B cells undergo class switch and affinity maturation, resulting in plasma cells secreting high-affinity IgG specific for the polysaccharide.

The development of genetic engineering holds great promise for the future of vaccinations. Already some of the subunit and toxoid vaccines that used to have to be purified from bulk cultures of the relevant organisms can be produced using recombinant DNA technology so that the relevant genes are introduced into bacteria or yeast that produce large quantities of the protein. However, other applications of DNA technology raise the prospect of developing vaccines that are safer, cheaper and more stable than many of the present vaccines. These aspects are all important, especially for the use of vaccines in countries where refrigerated storage conditions are not widely available. The new styles of vaccine being developed for trial include the following:

- **Recombinant vector vaccines.** These are vaccines in which the genes encoding important antigens for a pathogen are introduced into the genome of attenuated viruses or bacteria (Figure 10.3). An example would be to introduce the gene for a pathogenic antigen into the genome of the vaccinia virus, which previously was used for immunisation

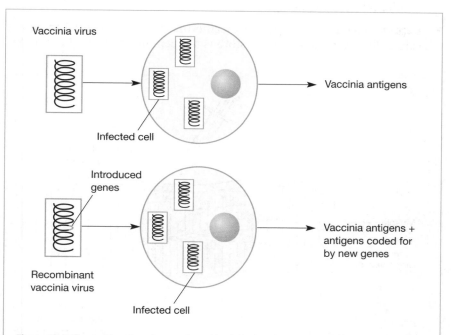

Figure 10.3 Recombinant vector vaccines. Vaccinia virus is an attenuated virus previously used as a vaccine for smallpox. Because it is a live virus it is infectious and stimulates antibody production, Th cells and cytotoxic CD8 T cells. By genetically modifying vaccinia virus it is possible to introduce genes coding for antigens of other pathogens. The virus directs the production of these antigens when it infects host cells and can therefore stimulate protective immune responses against the other pathogen.

against smallpox. The vaccinia virus, in addition to directing the expression of its own antigens, would also cause expression of the pathogen's antigen and stimulate immunity against smallpox (which is not necessary because natural infection with smallpox should not occur) and the pathogen. Because the vaccinia virus can infect cells it should stimulate good antibody responses and cell-mediated immunity against the pathogen's antigen.

- **DNA vaccines.** One fairly recent finding is that when muscle cells are exposed to DNA, they can take up the DNA and express the proteins coded for by the DNA. This happens *in vivo* and provides a mechanism for inducing the production of a wide variety of proteins including pathogenic antigens. In practice the DNA is in the form of a plasmid (Figure 10.4). This means that promoters can be introduced into the plasmid, which causes high production of the pathogen's protein. Experimental (DNA) plasmid vaccines induce both antibody- and cell-mediated immunity and have been shown to provide immunity against infection with the pathogen from which the antigen DNA was derived. Clinical trials are under way with several DNA vaccines. One major advantage of DNA vaccines is that DNA is very stable and does not require refrigerated storage.

- **Peptide vaccines.** The type of vaccine that has the potential to be the safest, cheapest and easiest to store is the peptide vaccine. This consists

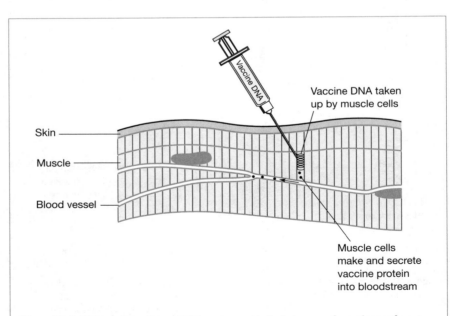

Figure 10.4 DNA vaccines. Plasmid DNA engineered to include genes for antigens of pathogens is taken up by muscle cells after intramuscular injection. The muscle cells express the antigen, which can stimulate protective immune responses.

Figure 10.5 Peptide vaccines. In this hypothetical example, a B cell epitope, a CD4 T cell epitope and a CD8 T cell epitope have been identified in a protein made by a pathogen. Knowing the amino acid sequences of these epitopes, it is possible to synthesise a peptide containing all three epitopes and this peptide can be used as a vaccine.

of a synthetic peptide that contains a CD4 T cell epitope and, depending on the pathogen, a B cell epitope and/or CD8 T cell epitope (Figure 10.5). They can be produced chemically and therefore, in bulk, cheaply and, because they do not involve DNA or the inactivation of toxins, there should be no risk of accidental exposure to toxins or virulent organisms. The main drawbacks to peptide vaccines have been identifying appropriate epitopes and making the peptides immunogenic. Their development has therefore been slower than originally hoped.

Overall vaccination has been a tremendously successful form of manipulating the immune system by stimulating specific immune responses against antigens on pathogens and has led to a significant reduction in mortality and morbidity.

10.3 Summary

- Immunological memory is a mechanism for making immune responses to a second, or subsequent, exposure to an antigen faster and bigger.

- Memory for antibody responses is based on memory B cells produced in germinal centres during a primary antibody response. Because memory B cells have already switched antibody class to IgG, IgA or IgE and have undergone affinity maturation, they are able to differentiate rapidly into high-affinity antibody-secreting plasma cells, providing a quicker response. There are also more memory B cells, providing a bigger memory response.

- Memory CD8 T cells are also present in higher numbers than prior to initial exposure to antigen and are able to differentiate more rapidly into Tcs, which also secrete more cytokines.

- Immunological memory provides the basis for vaccination and there are many different vaccines which are in use or being developed.

- For vaccines to be effective they must stimulate the right type of response and be safe. Ideally they should also be cheap, stable and easy to administer.

- Current vaccines are attenuated or heat-killed versions of the pathogen, subunit vaccines that are antigens from the pathogen, or toxoids that are inactivated toxins. They have provided protection against many illnesses with considerable mortality or morbidity but there are many infections for which no vaccines are available.

- Vaccines under development or being introduced for clinical use are recombinant vector vaccines, DNA vaccines or peptide vaccines. They have the advantages that they may be safer, cheaper and easier to store than many current vaccines.

CHAPTER ELEVEN

Lymphocyte development and immunological tolerance

Learning objectives

To learn how B and T cells develop from bone-marrow precursors. To know about the thymus and its role in T cell development. To understand the mechanisms of preventing the immune system reacting against self-antigens – self-tolerance.

Key topics

- Production of B cells
  - ○ Cellular aspects
  - ○ Tolerance
- Production of T cells
  - ○ Structure of the thymus
  - ○ Cellular aspects
  - ○ Positive and negative selection
- Peripheral tolerance in T cells
  - ○ Anergy
  - ○ Regulation
  - ○ Ignorance

11.1 Why must lymphocytes be produced continually?

The immune system generates lymphocytes with specificities for a huge range of antigens on any pathogen by randomly rearranging Ig and TcR genes during lymphocyte development. This means that following infection by a pathogen there will be some lymphocytes that are specific for antigens on the pathogen; these lymphocytes can be considered *useful* lymphocytes.

However, many lymphocytes will have antigen receptors that are specific for an antigen that the immune system will never encounter, although the immune system cannot know this. Through no fault of their own these lymphocytes are *useless* to the individual.

Since we have room for only a fixed number of lymphocytes the immune system will operate most efficiently if many useful lymphocytes are present at the expense of useless ones. The way this is achieved is to have a continual process of producing lymphocytes with a finite lifespan. If, during its lifespan, a lymphocyte encounters antigen for which its antigen receptors are specific, it is a useful lymphocyte and it will be stimulated to divide and become an effector or memory cell. If the lymphocyte does not meet its antigen within a certain time period it is deemed useless and dies; it can then be replaced by a new lymphocyte that has the potential to be useful. In this way useful lymphocytes are retained and useless ones are lost.

We are in a situation, then, where we continually produce lymphocytes. The random nature of Ig/TcR gene rearrangement, while being tremendously efficient at generating a huge variety of antigen specificities, means that lymphocytes can be generated with specificity for *any* antigen, including self-antigens. Lymphocytes with specificity for self-antigen are potentially *dangerous* and must be prevented from reacting against self-antigens. The process of stopping lymphocytes reacting against self is known as **self-tolerance** and occurs partially through mechanisms that operate during B and T cell production and partially through control of the activity of mature lymphocytes.

11.2 The production of lymphocytes: lymphopoiesis

Lymphocytes are bone-marrow-derived cells and are produced originally from a pluripotential stem cell. This is a cell in the bone marrow that is ultimately responsible for maintaining numbers of all haematopoietic cells. It is capable of division, and thereby maintaining cell numbers, and can differentiate into any of the different haematopoietic cells (see Figure 11.1).

Pluripotential stem cells produce lymphocytes through processes of cell division and differentiation. As the stem cells differentiate they become more and more specialised. The pluripotential stem cell, as its name implies, is capable of differentiating into any of the blood cell types. In the production of lymphocytes some of the pluripotential stem cells become lymphopoietic stem cells, which are also called common lymphocyte progenitors. These cells can become B lymphocytes or T lymphocytes.

One major factor in determining whether a precursor cell becomes a B cell or a T cell is the anatomical site in which it develops. B cells develop in the bone marrow and T cells develop in the thymus (Figure 11.1). Another difference is that T cells recognise antigen in association with MHC (see Chapter 4) whereas B cells recognise free antigen; this imposes additional requirements on T cell development that are not seen with B cells, as discussed below.

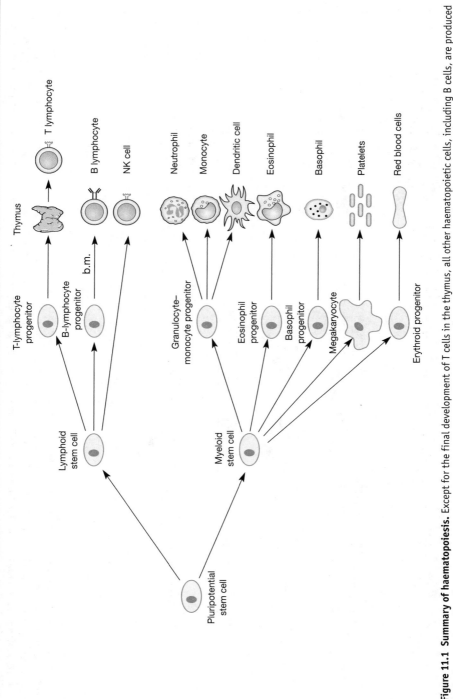

Figure 11.1 Summary of haematopoiesis. Except for the final development of T cells in the thymus, all other haematopoietic cells, including B cells, are produced in the bone marrow (b.m.).

Figure 11.2 In-frame and out-of-frame gene rearrangements during Ig or TcR gene rearrangement. ① In-frame rearrangements involve the loss of no nucleotides or multiples of three nucleotides (three, six, etc.), thereby maintaining the triplet codon sequence. ② The number of nucleotides lost in an out-of-frame rearrangement is not a multiple of three (one, two, four, five, seven, etc.), resulting in loss of the original triplet codon sequence. ③ The out-of-frame rearrangement can result in the creation of a stop codon. ④ The out-of-frame rearrangement results in the loss of one nucleotide in the rearranged-gene. This shifts the way the DNA is read in the constant region gene, creating new codons that result in a nonsense protein with no function.

During lymphocyte development, precursors of B or T lymphocytes undergo rearrangement of their antigen receptor genes. Each developing lymphocyte must successfully undergo at least three gene rearrangements. B cells must rearrange V to J in their Ig light chain gene plus V to D and D to J in their Ig heavy chain gene. Similarly T cells must rearrange V to J in their α-chain gene and V to D and D to J in their TcR β-chain gene. At each gene rearrangement there is only a one-in-three chance that the rearrangement will be in frame. An out-of-frame rearrangement acts like a mutation and can create a stop codon or a nucleotide sequence that codes for a nonsense protein (Figure 11.2). In this case the developing lymphocyte will not be able to make the protein required to form a receptor for antigen and will be non-functional. It is most efficient to detect as early as possible whether a developing lymphocyte has failed to rearrange its antigen receptor genes and terminate the development of the lymphocyte if it has been unsuccessful. Therefore B and T cells rearrange their Ig H or TcRβ genes first since this involves two rearrangements and is most likely not to be successful. The lymphocytes then have to go through a checkpoint where they are tested to see whether they have successfully rearranged the first gene (Box 11.1). Only those cells that have successfully rearranged their first gene are allowed to continue their development. This stops any more energy being wasted on producing non-functional lymphocytes.

In addition to checking that a lymphocyte has successfully rearranged one of each of its receptor genes it is important to ensure that lymphocytes do not rearrange both copies of a receptor gene. Remember that lymphocytes have two copies of each receptor gene, one inherited from the mother and one inherited from the father. If the first receptor gene is not successfully rearranged the lymphocyte has another chance and can rearrange the second gene. However, if the first receptor gene is successfully rearranged, the lymphocyte must stop the second copy of the gene from rearranging. The process of stopping the second receptor gene rearranging is called **allelic exclusion** (Figure 11.3 and Box 11.2) and is very important. If allelic exclusion did not occur a lymphocyte could successfully rearrange both copies of a receptor gene and make two proteins. Each protein could form part of an antigen receptor but the two receptors would have different antigen specificity. Having lymphocytes with two receptors with different antigen specificities would make control of antigenic-specific responses chaotic.

11.3 Production of B lymphocytes

B lymphocytes are produced in the bone marrow. The first stage in B cell development is a rearrangement of D to J in the Ig heavy chain gene, involving D rearranging to J (Figure 11.4). Cells that successfully rearrange D to J are known as **pro-B cells**. These cells also express a molecule called CD19 on their cell surface, which identifies them as cells of the B lymphocyte

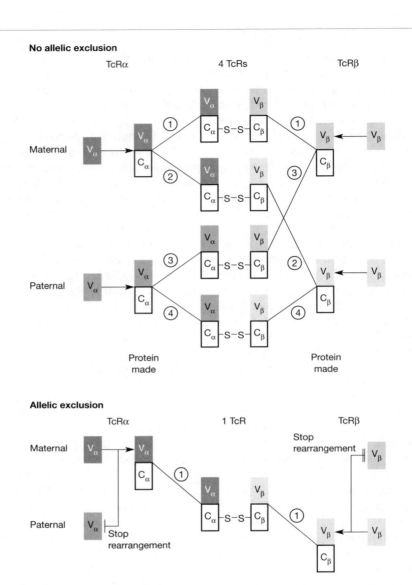

Figure 11.3 Allelic exclusion prevents lymphocytes expressing more than one receptor antigen specificity. The figure shows the situation how TcR expression would occur with and without allelic exclusion. Without allelic exclusion, both maternal and paternal Vα genes and Vβ genes may successfully rearrange, resulting in the production of two different α-chains and two different β-chains. The maternal Vα can associate with either the maternal or the paternal Vβ and the paternal Vα can associate with either the maternal or the paternal Vβ, giving four different TcRs with four different antigen specificities. The numbers in circles show pairings of Vα and Vβ. Allelic exclusion stops the second copy of the gene rearranging if the first copy rearranges successfully, therefore ensuring each T cell makes only one α-chain and one β-chain and therefore one TcR with one antigen specificity. For B cells the situation is even more complicated because a B cell could theoretically generate two different H-chains and four different L-chains (2κ and 2λ), giving a total of eight different receptor specificities.

BOX 11.1: CHECKPOINTS IN B AND T CELL DEVELOPMENT

If a B cell successfully rearranges one of its heavy chain genes it will begin to make heavy chain protein. The heavy chain protein is expressed on the cell surface in association with a non-rearranged Ig light chain made up of a VpreB variable segment and a λ5 constant segment (see figure below). The expression of this receptor is essential for the future differentiation of the B cell and provides an important checkpoint. The exact way in which it works is not known but the receptor presumably engages some as yet unknown ligand, which results in the cell receiving a survival signal. Cells that have not successfully rearranged their H-chain genes will not be able to express the receptor and therefore will not continue the differentiation process and will die instead. This VpreBλ5 light chain is expressed only at this stage of B-cell development and has been called a 'surrogate' or 'pseudo' light chain because it is not involved in recognition of conventional antigen.

In a similar manner T cells that have successfully rearranged their TcR β-chain gene will express the β-chain on the cell surface in association with a pre-T cell α-gene product and CD3. Like B cells, the expression of this receptor is essential for the continued development of the thymocyte, although again the mechanism is not known. Thymocytes that do not successfully rearrange their β-chain genes cannot express the receptor and die by apoptosis.

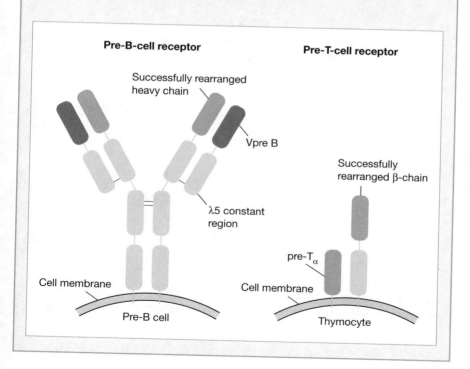

lineage. The next stage of B cell development is rearrangement of an Ig heavy chain V segment to the rearranged DJ gene. Cells that have successfully completed both the D to J and the V to DJ rearrangements are allowed to continue their development and are called **pre-B cells**. Cells that do not successfully rearrange their heavy chain genes die (see Box 11.1).

The next stage of B cell development is rearrangement of the Ig light chain genes. If the rearrangement of V to J is successful the cell can make an Ig light chain that associates with the Ig heavy chain and the B lymphocyte will express IgM on its cell surface. This cell is now known as an **immature** B cell. Finally the B cell will also express IgD on its surface, at which point it has completed the antigen-independent stage of maturation. This cell is called a 'naive' or 'virgin' B cell because it has never experienced antigen.

Naive B cells are exported to the peripheral lymphoid tissue where they will circulate through blood and lymphoid tissue and wait to encounter the antigen for which their Ig is specific. If they encounter antigen in the presence of appropriate T cell help they will undergo antigen-dependent maturation and differentiate into plasma or memory cells. Most B cells that do not encounter antigen within a week of leaving the bone marrow die.

Figure 11.4 B cell development. The pro-B cell expresses CD19 on its cell surface and rearranges the IgH D gene to the J gene. The pre-B cell rearranges its V gene to the DJ and expresses the Ig μ-chain in its cytoplasm. Immature B cells express IgM but not IgD on their cell surface while mature B cells express both IgM and IgD.

BOX 11.2: ALLELIC EXCLUSION

Allelic exclusion is a mechanism to stop lymphocytes expressing two different anti-gen receptors. The situation is relatively straightforward in B cells but not so well understood in T cells.

The model for B cells fits with the available data. This model proposes that developing B cells in the bone marrow begin to rearrange *one* copy of their heavy chain genes. If the rearrangements (D to J and V to D) are successful the heavy chain that is made stops rearrangement of the other copy of the heavy chain gene. If the first heavy chain gene rearrangement is unsuccessful the cell will rearrange the second copy of the heavy chain gene. Assuming one of the heavy chain genes has successfully rearranged, the cell will then begin to rearrange one copy of the κ-chain genes. If this is successful the κ-chain produced can associate with the heavy chain to form a complete antibody. This antibody stops rearrangement of the other κ gene and both λ genes. If the rearrangement is unsuccessful the cell will carry on re-arranging the other κ gene and λ genes until it has tried to rearrange all four genes. At any stage, if a light chain gene rearrangement is successful the antibody produced as a consequence will stop rearrangement of any un-rearranged light chain genes.

A similar situation is thought to occur to some extent with T cells. A successful rearrangement of the first TcR β-chain gene stops rearrangement of the second heavy chain gene. However, the successful rearrangement of the first α-chain gene does not necessarily stop rearrangement of the second α-chain gene and some T cells can be produced that have two TcR receptors. Both the receptors have the same β-chain but they are associated with different α-chains.

11.3.1 Self-tolerance in B cells

The two main mechanisms of preventing B cells from reacting against self-antigens are clonal deletion and clonal anergy. The environment in the bone marrow where B cells develop is usually sterile and devoid of foreign antigens. Therefore the only antigen that developing B cells could encounter is self-antigen. If immature B cells, expressing IgM but not IgD on their surface, recognise antigen through their IgM they undergo one of two fates (Figure 11.5):

- **Clonal deletion.** In this situation the developing B cell recognising self-antigen is induced to die by apoptosis and obviously will not develop into a mature B cell. For this reason the process is known as clonal deletion.

- **Clonal anergy.** Alternatively the developing B cell may be rendered unresponsive. These unresponsive cells are now unable to respond to their particular antigen and are called anergic.

Functionally speaking, clonal deletion and clonal anergy are the same. With clonal deletion developing self-reactive B cells are eliminated and with clonal anergy the self-reactive B cell is still there but can no longer respond to antigen. In neither situation will self-reactive antibody be produced.

It is not clear why both mechanisms exist but the nature of the antigen seems to affect whether cells are deleted or anergised. Experiments have

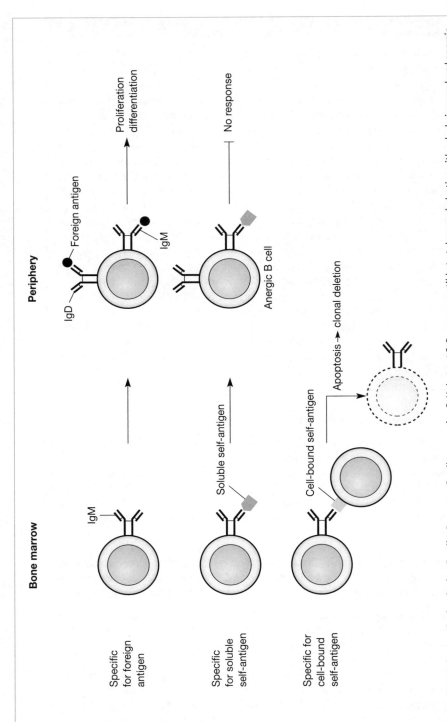

Figure 11.5 Tolerance in developing B cells. Immature B cells expressing IgM, but not IgD, are susceptible to tolerance induction either by being rendered anergic if their membrane IgM binds soluble self-antigen, or by apoptosis if their membrane IgM binds cell-bound self-antigen in the bone marrow.

shown that antigens on the surface of other cells induce clonal deletion but that soluble antigens induce anergy.

11.4 Production of T lymphocytes

Although the fundamental requirements for differentiation of B and T lymphocytes are similar, the situation with T cells is more complex than with B cells for two reasons. All B cells produced in the bone marrow are the same except for the specificity of their surface IgM and IgD. By contrast three types of T cell are produced in the thymus from the same type of precursor cell (Figure 11.6). Two of these cell types, the CD4 and CD8 T cells, express an α/β TcR, and the third type bears a different receptor for antigen, the γ/δ TcR. The rest of this section will cover the production of α/β TcR-bearing T cells; the production of γ/δ T cells is covered in Box 11.3. All of these cell types are produced in the thymus. The second reason why T cell development is more complicated is that T cells recognise antigen in association with MHC. T cells must be selected in each individual to recognise antigen in association with that individual's own MHC, a process known as thymic education (see Section 11.4.4).

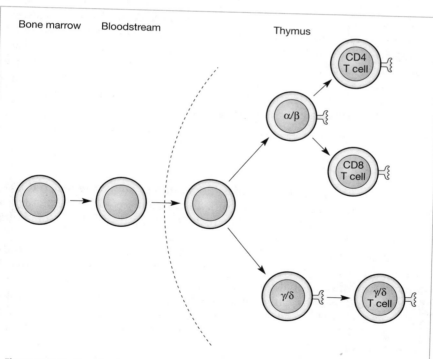

Figure 11.6 T cell production in the thymus. T cell progenitors enter the thymus from the blood and can develop into CD4 or CD8 T cells bearing the α/β TcR or T cells expressing the γ/δ TcR.

BOX 11.3: γ/δ T CELLS

A small population of T cells exists that express a different antigen receptor from the conventional α/β receptor expressed by CD4 and CD8 T cells. This receptor is composed of a γ-chain and δ-chain that pair to form a dimer that is expressed on the cell surface in association with CD3. The γ and δ genes are similarly arranged to the Ig and TcR genes, consisting of V, J and C gene segments plus a D segment in the case of the δ gene. These genes undergo rearrangement in the same way as the Ig and TcR genes and can generate a large diversity of antigen-specific receptors (see Chapter 5).

Less than 5% of T cells are γ/δ T cells and much about them is not clear. They are found at the highest numbers in epithelial sites such as the epidermis and epithelium of the gut, intestine and vagina. Most of the γ/δ T cells in one site express the same Vγ and Vδ gene segments and many have the same Jγ, Jδ and Dδ segments with little or no junctional diversity. This suggests that their recognition of antigen is limited. Different tissues will have γ/δ T cells using different V, D and J segments from those in another tissue. The way in which γ/δ T cells recognise antigen differs from α/β T cells in that it is not MHC-restricted. Some γ/δ T cells still need antigen to be somehow associated with MHC, but the allele of the MHC does not matter. Other γ/δ T cells recognise antigen without any apparent involvement of MHC.

The function of γ/δ T cells is also not clear. *In vitro* they are capable of killing tumour and other cells and secrete many cytokines. Cytotoxicity and cytokine secretion occur rapidly *in vitro* without any apparent need to stimulate the post-antigenic stimulation differentiation required by B and T cells (although it cannot be excluded that the cells are not continuously being stimulated by antigen *in vivo*). Because of their immediate response and location primarily in epithelial sites it has been suggested that they perform a sentinel role, waiting in the epithelium to provide an immediate response to infection.

11.4.1 The thymus

The thymus is located in the mediastinum and lies above the heart. It has a two-lobed structure and each lobe is divided into many lobules separated by trabeculae of connective tissue (Figure 11.7 and Plate 12). Each lobule has two sections that can be easily distinguished histologically. The outer area stains more densely with haematoxylin and eosin (H & E) and is called the cortex. The less densely staining inner section is called the medulla. It should be mentioned that there is no apparent anatomical barrier between the cortex and medulla, although particulate material injected intravenously is able to access the medulla but not the cortex. The thymus reaches its peak size around puberty, after which it shrinks or involutes. In adults it is composed mostly of fat cells and connective tissue, although it produces new T cells well into old age.

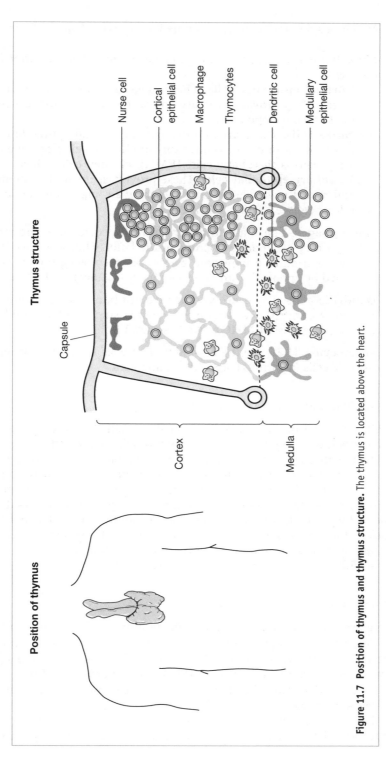

Figure 11.7 Position of thymus and thymus structure. The thymus is located above the heart.

Numerous different cell types can be found in the thymus:

- **Epithelial cells.** Three types of epithelial cell have been described in the thymus:
  (i) **Cortical epithelial cells.** These make up a mesh throughout the cortex. They provide structure and secrete factors that are essential for T cell development.
  (ii) **Nurse cells.** These are another type of cortical epithelial cell and can be seen in close contact surrounding developing thymocytes. They express class I and class II MHC on their cell surface.
  (iii) **Medullary epithelial cells.** These provide structure for the medulla and recent evidence suggests they play an important part in T cell tolerance (see Section 11.5).

- **Macrophages.** These are found in both the cortex and the medulla, although they are more numerous in the latter. They are class II MHC negative and play an important role in phagocytosing thymocytes that have died by apoptosis during development (see Section 11.4.3).

- **Dendritic cells.** These are another type of bone-marrow-derived cell related to dendritic cells found in most tissues and lymphoid tissue. They express class II MHC as well as class I MHC.

- **Thymocytes.** These are the most abundant cells in the thymus and consist of T cells in various stages of development.

11.4.2 Anatomical aspects of T cell development

T cells develop from bone-marrow precursors called prothymocytes. These precursors travel from the bone marrow through the bloodstream and enter the thymus through venules at the cortico-medullary junction (Figure 11.8). The most immature thymocytes are found in the subcapsular region of the cortex. These cells are rapidly dividing and give rise to large numbers of thymocytes. As the developing thymocytes carry on with the maturation process they stop dividing and move deeper into the cortex. Once they have finished their development T cells either leave the thymus via venules or lymphatic vessels located at the cortico-medullary junction, or migrate to the medulla eventually leaving by the same routes.

11.4.3 Differentiation of thymocytes

There are two vital aspects to the production of T cells. The first of these is the differentiation of prothymocytes so that they acquire the function associated with mature CD4 or CD8 T cells. This involves altering gene expression and can be followed by testing for the presence of new proteins in the cells. Secondly the thymocyte must rearrange its TcR genes and express a receptor for antigen/MHC.

The differentiation of thymocytes can be followed by examining the expression of TcR and certain CD molecules on the developing thymocytes

Figure 11.8 T cell development in the thymus. Cells enter the thymus from venules at the cortico-medullary junction ① and travel to the subcapsular area of the outer cortex ②. These cells, which do not express CD4 or CD8, divide rapidly for a while and then stop dividing and migrate into the cortex ③, where they express both CD4 and CD8. As the cells traverse from the cortex to the medulla they stop expressing CD4 or CD8 ④ and leave the medulla as naive T cells through blood vessels at the cortico-medullary junction ⑤.

(Figure 11.8). The most immature thymocytes that are found under thymic capsule express CD2 but not CD4 or CD8 and are therefore called CD4⁻CD8⁻ thymocytes, or double-negative thymocytes. At the end of the double-negative stage of differentiation the thymocytes rearrange their TcR β genes. Those that successfully rearrange one of their TcR β genes and pass the checkpoint (see Box 11.1) rearrange their TcR α genes and, if this is successful, the cells express low levels of the TcR on their cell surface. These small resting thymocytes also express both CD4 and CD8 on their surface and are called double-positive thymocytes; they represent about 85% of all thymocytes and undergo positive and negative selection (see below). Finally, when the thymocytes cells arrive at the cortico-medullary junction they stop expressing either CD4 or CD8 and become mature single positive thymocytes expressing either CD4 or CD8, which either go to the medulla or leave the thymus.

11.4.4 Positive and negative selection of thymocytes: thymic education

T cells recognise antigen in an MHC-restricted manner. Some of the TcR binds to MHC and some binds to the peptide (see Section 4.4). Because the rearrangement of TcR genes is random, T cells can be produced with four potential types of specificity in terms of MHC and antigenic peptide. The TcR could be specific for:

1. Foreign antigenic peptide + self-MHC.
2. Self-antigenic peptide + self-MHC.
3. Foreign antigenic peptide + foreign MHC.
4. Self-antigenic peptide + foreign MHC.

Of the four specificity patterns, only cells that are specific for foreign antigenic peptides plus self-MHC (pattern 1) will be useful at recognising antigens derived from pathogens. Cells with specificity for self-antigenic peptide plus self-MHC (pattern 2) are potentially damaging and could cause an immune response against the body's own antigens. Cells that can only recognise antigen in association with foreign MHC (patterns 3 and 4) will be useless in that particular individual. Therefore during T cell development there is the requirement (i) to select T cells that are self-MHC-restricted, called positive selection and (ii) to prevent the production of T cells with specificity for self-antigenic peptide and self-MHC, called negative selection. These are accomplished by special mechanisms that operate during T cell development in the thymus.

Positive selection
The term 'positive selection' is used to describe the processes by which thymocytes that are specific for self-MHC are distinguished from those that are not specific for self-MHC (Figure 11.9). During thymocyte development, double-positive thymocytes, which have rearranged their TcR genes and express a receptor for antigen, encounter class I or class II MHC on the surface of thymic cortical epithelial cells. Thymocytes whose TcR can bind the self-MHC expressed on the epithelial cells receive survival signals and proceed to the next stage of differentiation. Thymocytes whose TcRs cannot bind with self-MHC do not receive the survival signals and die by apoptosis. The apoptotic cells are rapidly phagocytosed by thymic macrophages. This process means that only thymocytes whose TcR can bind self-MHC survive, but it does not distinguish between thymocytes that are specific for self-antigenic peptide + self-MHC and thymocytes that are specific for foreign antigenic peptide + self-MHC. Therefore a process is necessary to eliminate thymocytes with specificity for self-antigen; this process is called negative selection.

Negative selection
The area of the thymus at the cortico-medullary junction contains many dendritic cells expressing both class I and class II MHC, which can present peptides derived from self-antigens to the developing thymocytes (Figure 11.9). The developing thymocytes encounter these dendritic cells and if they react with self-antigenic peptide plus self-MHC with high enough affinity they are stimulated to undergo apoptosis and are phagocytosed. Thymocytes that do not react with self-antigenic peptide + self-MHC are allowed to finish the maturation process and emerge from the thymus as mature CD4 or CD8 T cells.

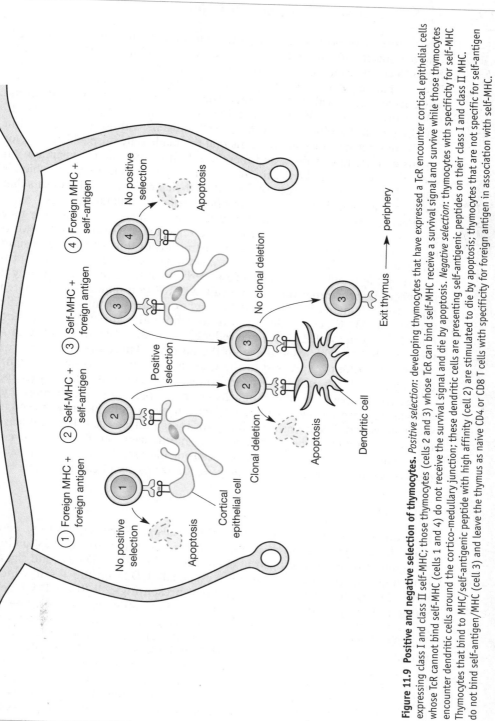

Figure 11.9 Positive and negative selection of thymocytes. *Positive selection:* developing thymocytes that have expressed a TcR encounter cortical epithelial cells expressing class I and class II self-MHC; those thymocytes (cells 2 and 3) whose TcR can bind self-MHC receive a survival signal and survive while those thymocytes whose TcR cannot bind self-MHC (cells 1 and 4) do not receive the survival signal and die by apoptosis. *Negative selection:* thymocytes with specificity for self-MHC encounter dendritic cells around the cortico-medullary junction; these dendritic cells are presenting self-antigenic peptides on their class I and class II MHC. Thymocytes that bind to MHC/self-antigenic peptide with high affinity (cell 2) are stimulated to die by apoptosis; thymocytes that are not specific for self-antigen do not bind self-antigen/MHC (cell 3) and leave the thymus as naïve CD4 or CD8 T cells with specificity for foreign antigen in association with self-MHC.

11.5 Peripheral tolerance in T cells

The processes of positive and negative selection result in peripheral T cells that are specific for foreign peptide in association with self-MHC and therefore select for useful T cells and eliminate useless or potentially damaging T cells. However, the process is not 100% efficient because not all self-antigens are expressed in the thymus. The self-antigens that thymic dendritic cells express on their MHC must either be made in the thymus or travel to the thymus in the bloodstream. Many self-antigens will not be expressed in the thymus because they are proteins that are made only in specific tissues or organs and are not secreted and therefore do not enter the bloodstream. An example of this would be the thyroid-stimulating hormone receptor, which is only expressed on the surface of thyroid epithelial cells. The self-antigens that are expressed outside the thymus have been called peripheral antigens.

Recent evidence has indicated that thymic medullary epithelial cells may express proteins that would normally be regarded as tissue-specific extra-thymic proteins. One of the first demonstrations of this was that of insulin expression in the thymus; previously it was thought that insulin was only produced by the β cells in the islets of Langerhans of the pancreas. Recently it has been demonstrated that a gene called AIRE (autoimmunity associated regulatory element) is important for thymic expression of tissue-specific genes. The AIRE gene codes for a transcriptional regulatory protein and mice lacking the AIRE gene have reduced expression of extra-thymic proteins by their thymic medullary epithelial cells. Therefore there may be more self-proteins expressed in the thymus than was previously thought, although not all self-proteins of an individual will be expressed in the thymus.

Because many peripheral antigens are not expressed in the thymus, thymocytes that are specific for these self-antigens will not be clonally deleted in the thymus and T cells with specificity for these antigens can therefore leave the thymus and enter the periphery. Other mechanisms must exist to prevent these autoantigen-specific T cells from causing disease. Because these mechanisms operate outside the thymus they are known as peripheral tolerance.

11.5.1 Clonal anergy

It was described in Chapter 7 how CD4 T cells need two signals to be activated by antigen. The first of these signals is through the TcR following recognition of specific antigen in association with MHC. The second signal is through co-stimulatory molecules, especially CD28 on the T cell, which binds to CD80 or CD86 on the antigen-presenting cell. If a CD4 T cell recognises antigen/MHC without receiving the co-stimulus, not only does it fail to be activated but it is made **anergic** (Figure 11.10). This means that the CD4 T cell is rendered incapable of responding to antigen and is a particularly important mechanism for peripheral tolerance in CD4 T cells.

Most parenchymal cells (i.e. cells, such as hepatocytes in the liver, that make up different tissues) do not express class II MHC. Therefore even if they have peripheral antigen they cannot present them to CD4 T cells so there is no risk of stimulating autoimmune CD4 T cells. Many cell types can be induced to express class II MHC – e.g. by IFN-γ – so that during an

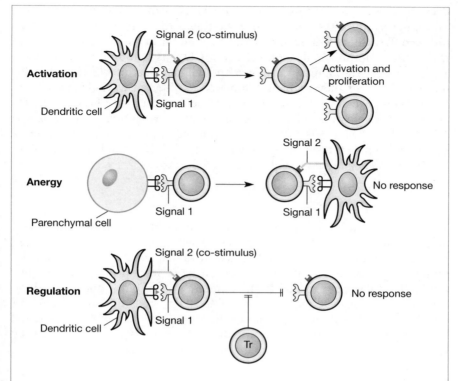

Figure 11.10 Peripheral tolerance in T cells. T cells need two signals to be activated, one signal from the TcR-binding antigen/MHC and a co-stimulus provided by the antigen-presenting cell (top). If the T cell receives signal 1 alone it is rendered anergic and cannot respond to a subsequent encounter with the same antigen even if it receives the co-stimulus (middle). Regulatory cells (Tr) are able to stop self-reactive CD4 T cells responding even if the CD4 T cell receives the two signals required for activation.

immune response to a virus in a tissue the parenchymal cells could be induced to express class II MHC. These cells now have the capacity to present peripheral self-antigen to CD4 T cells. However, these parenchymal cells do not express co-stimulatory molecules and therefore cannot deliver co-stimulatory signals. CD4 T cells that recognise a tissue-specific self-antigen being presented by class II MHC on a parenchymal cell will therefore be anergised, so preventing the activation of the autoreactive cells.

11.5.2 Active regulation

There is evidence that cells exist in the periphery that are able to prevent autospecific T cells from reacting against self-antigens (Figure 11.10). These cells are CD4 T cells and have been called regulatory T cells, Tregs or Trs.

The situation with regulatory T cells is somewhat confusing, mainly because there appear to be a number of different types and it is not clear how they are related to each other. Some Trs have been called natural Trs because they are present without any deliberate exposure to antigen. Other Trs appear only after deliberate induction with antigen and are called induced Trs. Some Trs have CD25, the α-chain of the IL-2 receptor, on their surface but others do not. However, CD25 is found on activated CD4 T cells that are responding to antigen and will develop into Th cells. Finally, different Trs exert their regulatory function in different ways (Figure 11.11). Some Trs (especially natural Trs) require cell contact with either the CD4 T cell that is being regulated or the APC that is presenting antigen/class II MHC to the CD4 T cell. Other Trs do not require cell contact but secrete cytokines such as transforming growth factor-β (TGF-β) and/or interleukin-10 (IL-10), which are powerful immunosuppressants.

11.5.3 Clonal ignorance

The term 'clonal ignorance' is used to describe the situation where self-antigen-specific T cells fail to recognise the self-antigen because it is present in too low concentrations or because it is sequestered away from the immune system. An example of the latter is lens protein in the eye. Normally this is not exposed to the immune system but if one eye is damaged (e.g. by a squash ball) in some cases CD4 T cells that are specific for the lens protein and are not tolerant become activated and react against both the damaged eye and the undamaged eye. This condition is called reactive sympathetic opthalmia and can lead to blindness if untreated.

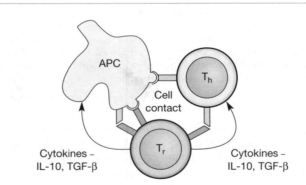

Figure 11.11 Regulatory T cells (Trs) can inhibit the proliferation and development of helper CD4 T cells (Ths) through cell–cell contact with the CD4 T cell or the APC that is stimulating the CD4 T cell. Trs can also inhibit CD4 T cells or APCs by the secretion of soluble factors (cytokines) such as IL-10 and TGF-β.

11.6 Summary

- Lymphocytes are produced continually – those that are stimulated by antigen (e.g. on an infectious organism) become effector or memory cells and those that do not see their specific antigen die in a determined time.

- B lymphocytes are produced in the bone marrow where they differentiate from lymphopoietic stem cells.

- Initially developing B cells rearrange their Ig heavy chain genes. Those that do this successfully then rearrange their light chain genes; those that are unsuccessful die.

- Immature B cells express IgM but not IgD. If these cells have specificity for self-antigen and bind that self-antigen in the bone marrow they undergo apoptosis (clonal deletion) or are rendered non-functional (clonal anergy). These processes help stop B cells responding against self-antigen, a process called self-tolerance.

- B cells finish their maturation by expressing both IgM and IgD on their surface. They leave the bone marrow and re-circulate through the blood and lymphoid tissue.

- T cells develop from immature precursors in the thymus, a bi-lobed organ in the mediastinum.

- Developing thymocytes rearrange their TcR β-chain gene and those that do so successfully go on to rearrange their TcR α-chain. Cells not successfully rearranging their TcR β-chain gene die.

- Thymocytes then express both CD4 and CD8 and an α/β TcR. They undergo positive selection where only those thymocytes with specificity for self-MHC survive.

- Surviving thymocytes then undergo negative selection where any cells that recognise self-antigen plus self-MHC die by apoptosis. Only T cells that recognise self-MHC and do not encounter self-antigen in the thymus leave the thymus as either CD4 or CD8 naive T cells.

- Because not all self-antigens are expressed in the thymus, mechanisms of peripheral tolerance exist. These include clonal anergy, clonal ignorance and active regulation.

CHAPTER TWELVE

Autoimmune diseases

Learning objectives

To understand what is meant by an autoimmune disease. To know about the spectrum and nature of autoimmune diseases. To understand how autoimmune responses cause disease. To know that both genetics and the environment contribute to the development of autoimmunity. To understand how tolerance may be lost.

Key topics

- Definition of autoimmune diseases and clinical burden
- Classification of autoimmune diseases
- Immunological features of autoimmune disease
  - Antibody responses
  - T cell responses
- Aetiology of autoimmune diseases
  - Genetic factors
  - Environmental factors
- Loss of tolerance

12.1 Definition and terminology of autoimmune disease

Autoimmune diseases are diseases that involve an immune response against one or more self-antigens. These self-antigens are usually proteins that constitute part of the body; less often they are carbohydrates, lipids or DNA. The self-antigens that the immune system responds to in an autoimmune disease are called **autoantigens** and the immune response against an autoantigen is called an **autoimmune response**.

The first disease to be identified as being autoimmune in origin was Hashimoto's thyroiditis in the mid-1950s. In this disease antibodies are produced against thyroglobulin and other thyroid-associated antigens. Antibodies against self-antigens are called **autoantibodies** and lymphocytes whose antigen receptors are specific for a self-antigen are called **autoreactive cells**. Nearly all autoimmune diseases involve the production of autoreactive CD4 T cells, which are also called **autoreactive Th cells**. Depending on the autoimmune disease there may also be the production of autoreactive B cells and/or autoreactive CD8 T cells. In Hashimoto's thyroiditis there is also an extensive infiltrate of lymphocytes and monocytes into the thyroid (Figure 12.1). The result of anti-thyroid antibody production and the thyroid infiltrate is the destruction of a large proportion of the thyroid glandular tissue, leading to underfunctioning of the thyroid, or hypothyroidism.

Hashimoto's thyroiditis demonstrates another feature of autoimmune diseases – the occurrence of **primary pathology** and **secondary pathology**. Primary pathology is the direct consequence of the autoimmune response; in the case of Hashimoto's thyroiditis this is destruction of thyroid tissue. The secondary pathology is a consequence of the altered tissue function caused by the primary pathology. Again, using Hashimoto's thyroiditis as an example, the secondary pathology due to hypothyroidism differs depending on the severity of the disease and the age at which it occurs. In infants it can cause cretinism, which is the retardation of physical and mental growth. In adults, some of the symptoms are fatigue, intolerance of the cold, dry skin and, in severe cases, mental impairment.

12.2 Spectrum and clinical burden of autoimmune diseases

The number of different autoimmune diseases is huge and essentially every tissue and organ can be affected by an autoimmune disease. Table 12.1 is a partial list of autoimmune diseases. Some of these diseases, such as type I insulin-dependent diabetes mellitus (IDDM), rheumatoid arthritis and multiple sclerosis are familiar to most people; however, most autoimmune diseases are not that well known. The full range of autoimmune diseases is probably not yet appreciated and there is speculation that other diseases whose cause is at present unknown, such as some forms of mental illness and alopecia (hair loss), may be due to autoimmunity. Autoimmune diseases are generally chronic in nature and can be quite debilitating; many of them are fatal if not treated.

The number of people affected by all autoimmune diseases is about 5% of the population, so as a whole autoimmune diseases are quite common. The frequency of individual diseases varies considerably. Rheumatoid arthritis (RA) is the most common autoimmune disease, with an incidence of 1–2% of the population. Other autoimmune diseases are much rarer; pemphigus vulgaris, an autoimmune disease causing severe blistering of the skin, affects fewer than one in 100 000 people. It is not surprising that the more severe diseases with higher fatality rates usually have the lowest incidence. One

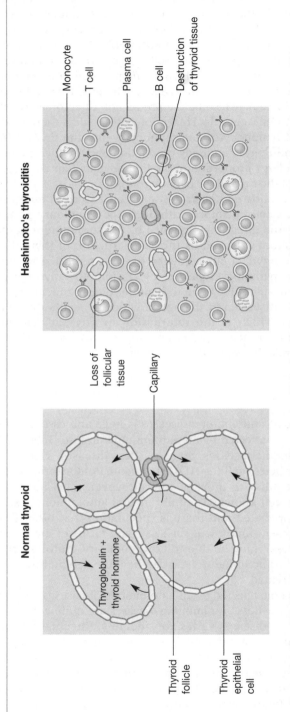

Normal thyroid

Loss of follicular tissue

Capillary

Thyroglobulin + thyroid hormone

Thyroid follicle

Thyroid epithelial cell

Hashimoto's thyroiditis

Monocyte

T cell

Plasma cell

B cell

Destruction of thyroid tissue

Figure 12.1 Thyroid infiltration and destruction in Hashimoto's thyroiditis. Normal thyroid consists of follicles containing thyroglobulin and thyroid hormone. The follicles are lined by thyroid epithelial cells. In Hashimoto's thyroiditis there is extensive infiltration of T cells, B cells, plasma cells and monocytes into the thyroid tissue, resulting in destruction of thyroid tissue and loss of thyroid hormone production.

Table 12.1 Spectrum of autoimmune diseases

Biological system affected	Disease	Main effects
Endocrine glands		
Thyroid	Hashimoto's thyroiditis	Thyroid destruction and underfunction
Thyroid	Grave's disease	Thyroid stimulation and overfunction
Islets of Langerhans (pancreas)	Insulin-dependent diabetes mellitus	Destruction of β cells (insulin-producing cells)
Adrenal gland	Addison's disease	Adrenal insufficiency
Haematopoietic system		
Red blood cells	Autoimmune haemolytic anaemia	Anaemia
Platelets	Autoimmune thrombocytopenia	Abnormal bleeding
Intrinsic factor (IF)	Pernicious anaemia	Autoantibody prevents absorption of vitamin B12
Nervous system		
Central nervous system	Multiple sclerosis	Progressive paralysis
Neuromuscular junction	Myasthenia gravis	Progressive muscle weakness
Skin		
Nuclear antigens	Scleroderma	Fibrosis of skin
Epidermal cell junctions	Pemphigus vulgaris	Severe blistering
Joints		
Synovium	Rheumatoid arthritis (RA)	Progressive destruction
Synovium	Systematic lupus erythromatosus (SLE)	Deformity
Kidney		
Basement membrane	Goodpasture's syndrome	Glomerulonephritis
Glomerulus	SLE	Glomerulonephritis

issue of   concern is that the incidence of autoimmune diseases has increased dramatically over the past 50 years or so, especially in industrialised countries.

12.2.1 Classification of autoimmune diseases

The wide spectrum of known autoimmune diseases has led to attempts to classify them. One distinction that became apparent between different autoimmune diseases was that some affect only one organ while others are more widespread. Diseases such as IDDM and thyroiditis affect the β cells of the islets of Langerhans of the pancreas and the thyroid specifically; no other organs or tissues are affected directly by the autoimmune response, although other tissues may be affected by the loss of insulin production or thyroid function caused by the autoimmune destruction of the tissue. Other autoimmune diseases affect many tissues, for example systemic lupus erythromatosus (SLE) is an autoimmune disease affecting the skin, kidneys and joints. Autoimmune diseases such as IDDM or thyroiditis, in which one organ is affected, are called **organ-specific**. Diseases such as SLE, where many tissues are affected, are called **systemic** or **non-organ-specific**. (See Box 12.1 for more details on the classification of autoimmune diseases.)

BOX 12.1: CLASSIFICATION OF AUTOIMMUNE DISEASES

The classification of autoimmune diseases as organ-specific or systemic is based on whether one tissue or multiple tissues are affected. This classification is not strictly based on the distribution of the autoantigens that are stimulating the autoimmune response although it was assumed that autoantigens in organ-specific diseases were present only in the tissues affected whereas the autoantigens in systemic autoimmunity were more widespread. In many cases this is true.

The classification of autoimmune diseases as organ-specific and non-organ-specific is based on the distribution of the autoantigens involved in the various autoimmune response. Autoimmune diseases in which the autoantigen is present only in the tissue affected are called organ-specific diseases. Again IDDM and thyroiditis would be classified organ-specific. Non-organ-specific autoimmune diseases are those in which the autoantigens are widespread and not confined to the tissue affected. The best characterised non-organ-specific autoimmune disease is SLE in which autoantibodies are produced against antigens such as DNA, histones and phospholipids, which are present in all nucleated cells.

It can be seen that there is considerable overlap between the two classification systems. However, there are problems with both classification systems and although they generally agree as to which diseases are organ-specific and which are systemic or non-organ-specific, there can be some conflicts. These can be illustrated by the autoimmune disease primary biliary cirrhosis (PBC). This is an autoimmune disease in which the bile ducts of the liver are destroyed, leading to loss of liver function. Because only the liver is affected by the autoimmune response PBC could be classified as an organ-specific autoimmune disease. However, the main autoantigens in PBC are mitochondrial antigens, which are present in almost all cells; this would classify the disease as non-organ-specific.

As more autoantigens are identified it is becoming clear that the autoantigens in other organ-specific autoimmune diseases are not always present just in the tissue affected. In IDDM many of the autoantigens are present in tissue outside the pancreas but these are not affected by the autoimmune response. Why the same autoantigen expressed in multiple tissue sites should stimulate autoimmune disease in one site and not others is one of the big puzzles in autoimmunity.

During an autoimmune response the immune system produces effector cells and molecules that attack particular parts of the body. These effectors may include antibody, cytotoxic CD8 T cells and effector CD4 T cells. Therefore, depending on the autoimmune disease, one or more of the serological and histological features associated with an attack by the immune system on a particular tissue or tissues will be seen.

- **Serum autoantibodies.** The serum of many patients with autoimmune disease contains autoantibodies. In the past these could sometimes be identified by immunofluorescence or immunocytochemical techniques using sections of relevant tissue (see Figure 12.2 and Plate 13). As more autoantigens are identified it is becoming easier to measure autoantibodies using enzyme-linked immunosorbent assay (ELISA) (see Figure 12.3).

- **Deposition of antibody and complement in affected tissue.** Another common feature of autoimmune diseases is the deposition of autoantibody in tissue. Autoantibody can bind directly to autoantigens in the tissue or it can be deposited in the form of immune complexes. Both of these situations can lead to complement fixation and activation, resulting in the generation of inflammatory responses. Deposition of antibody and complement can be detected by immunofluorescence or immunocytochemistry (see Figure 12.2 and Plate 13).

- **Infiltration of cells in the affected tissue.** In many autoimmune diseases a cellular infiltrate is seen in the affected organ or tissue (Plate 14). Typically the infiltrate will consist of lymphocytes and monocytes but other immune-related and inflammatory cells can sometimes be seen.

12.3.1 Antibody-mediated autoimmune diseases

In some autoimmune diseases autoantibody is the only or main autoimmune feature of the disease and the pathology can be explained by the actions of the autoantibody. Therefore it is easy to classify these as antibody-mediated autoimmune diseases. Different classes of autoantibody, with different functions, can be produced in autoimmune diseases and cause pathology in a number of ways:

- **Complement-mediated lysis.** Autoantibody binding to red blood cells can result in complement fixation and lysis of the red cells, leading to autoimmune haemolytic anaemia (Figure 12.4).

- **Opsonisation.** In autoimmune thrombocytopenia autoantibody binds to platelets and promotes their opsonisation by phagocytes in the liver and spleen (Figure 12.4). This leads to platelet deficiency (thrombocytopenia) and poor clotting.

Figure 12.2 Detection of autoantibody. (a) Autoantibody in serum can be detected by adding the serum to a section of normal tissue expressing the autoantigen ①. If autoantibodies are present in the serum they will bind to the autoantigens in the tissue section ②. In the immunofluorescence technique, a fluorescein-conjugated anti-human-Ig antibody is added to the section ③. This will bind to any autoantibody bound to the tissue section ④ and the antibody can be visualised by examining the section under an ultraviolet (UV) microscope ⑤. An alternative to immunofluorescence is immunocytochemistry. Instead of a fluorescein-conjugated antibody, an enzyme-linked anti-human-Ig antibody is used ⑥. The binding of the enzyme-linked antibody can be detected by adding a colourless substrate (S), which is turned into a coloured product (C) by the enzyme linked to the antibody ⑦. Immunofluorescence or immunocytochemistry can also be used to detect autoantibody that has been deposited in a tissue. In this case the fluorescein ⑧ or enzyme-linked ⑨ anti-human-Ig antibody is added to a tissue section from the person with the autoimmune disease without prior incubation with the patient's serum. The anti-human-Ig antibody will bind to autoantibody already bound to antigens in the

Figure 12.3 Detection of autoantigen by ELISA. Antigen is bound to a plastic well ①. Serum is added to the well and if autoantibody to the antigen is present it will bind to the antigen in the well ②. An enzyme-linked anti-human-IG antibody is added to the well and will bind to any autoantibody bound to the antigen ③. A colourless substrate (S) is added to the well and this will be converted into a coloured product (C) by the enzyme ④. The amount of coloured product produced can be measured using a spectrophotometer. The amount of product is proportional to the amount of enzyme, which in turn is proportional to the amount of autoantibody that bound to the autoantigen.

- **Inhibition of receptor function.** The disease myasthenia gravis involves the production of antibodies against the acetyl choline receptor, which is present on muscle fibres of the neuromuscular junction. The autoantibodies bind to the acetyl choline receptors and prevent the binding of acetyl choline released at the nerve endings. This blocks transmission of signals across the neuromuscular junction, leading to muscle weakness (Figure 12.5).

- **Stimulation of receptors.** For reasons that are not understood, in some cases antibodies against receptors can stimulate the receptor rather than block it. This happens in Grave's disease where autoantibodies are produced against the thyroid-stimulating hormone (TSH) receptor present on thyroid epithelial cells (Figure 12.5). Stimulation of these cells results in thyroid overactivity with the symptoms of hyper-thyroidism – nervousness, tiredness, weight loss despite a good appetite and proptosis (bulging of the eyes).

- **Blockage of biological function.** Some autoantibodies block the function of molecules other than receptors. Pernicious anaemia results from

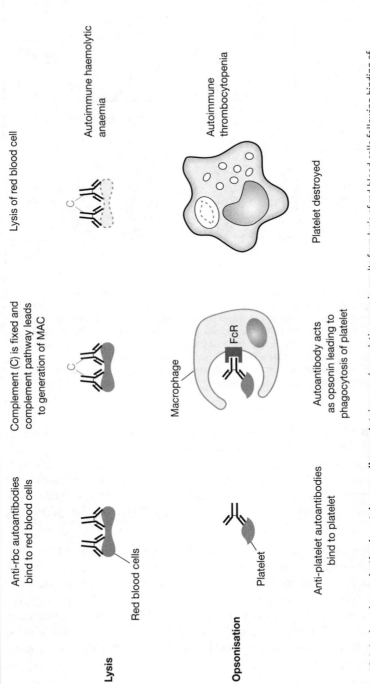

Lysis

Anti-rbc autoantibodies bind to red blood cells

Red blood cells

Complement (C) is fixed and complement pathway leads to generation of MAC

Lysis of red blood cell

Autoimmune haemolytic anaemia

Opsonisation

Anti-platelet autoantibodies bind to platelet

Platelet

Macrophage

FcR

Autoantibody acts as opsonin leading to phagocytosis of platelet

Platelet destroyed

Autoimmune thrombocytopenia

Figure 12.4 Lysis and opsonisation in autoimmune disease. Autoimmune haemolytic anaemia results from lysis of red blood cells following binding of autoantibodies to autoantigens on red blood cells and complement fixation. In autoimmune thrombocytopenia, binding of autoantibody to platelet autoantigens results in opsonisation and phagocytosis of the platelets and their destruction by phagocytes.

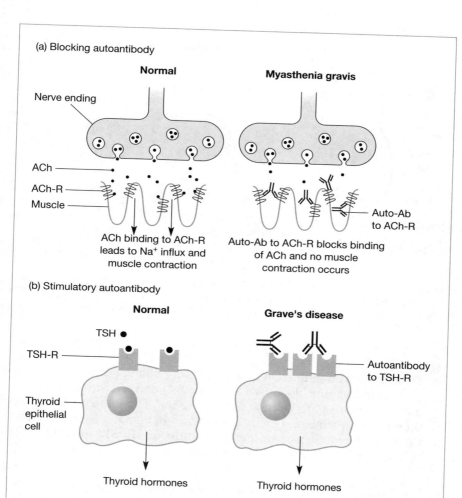

Figure 12.5 Antireceptor autoantibodies can block or stimulate function. In myasthenia gravis, autoantibodies are produced against the nicotinic acetyl-choline receptor present on muscle cells in neuromuscular junctions. The autoantibodies block the binding of acetyl-choline and prevent transmission of signals. The autoantibodies also cause internalisation and degradation of the ACh-R, reducing the number of receptors on the muscle cells. Autoantibodies against the thyroid-stimulating hormone receptor on thyroid epithelial cells stimulate the receptor rather than blocking it. This results in excessive secretion of thyroid hormones by the thyroid cells.

autoantibodies against intrinsic factor. Intrinsic factor is produced in the stomach and binds to vitamin B_{12}, enabling B_{12} to be absorbed from the intestine (Figure 12.6). The autoantibodies against intrinsic factor prevent the binding of B_{12} and hence its absorption, resulting in vitamin B_{12} deficiency. The vitamin deficiency leads to a lack of platelets and leukocytes and neurological changes.

Figure 12.6 Pernicious anaemia. Vitamin B_{12} is normally absorbed in the intestine following binding to intrinsic factor secreted by parietal cells of the stomach. Autoantibodies against intrinsic factor block the binding of vitamin B_{12} and it cannot be absorbed, resulting in vitamin B_{12} deficiency.

- **Deposition of immune complexes.** In SLE, immune complexes fail to be cleared from the blood (see Section 8.6.4) and are deposited in various sites such as the kidney, skin and joints. The immune complexes then fix complement, which leads to an inflammatory response (see Chapter 2) and damage to the affected tissues. In SLE this results in kidney and joint damage and characteristic skin rashes (Plate 15).

12.3.2 Autoimmune diseases with unknown causes of damage

In many autoimmune diseases it is not possible to blame the pathology simply on the action of antibody. In these diseases, which include IDDM, Hashimoto's thyroiditis and rheumatoid arthritis, there is extensive infiltration of immune-associated cells into the affected tissue and the production of autoantibodies (see above). The primary pathology in these diseases is damage to, or destruction of, part of the tissue, but the exact cause of the damage is hard to determine. Often the cellular infiltrate consists of CD4 and CD8 T lymphocytes, B cells, monocytes/macrophages and other inflammatory cells. Therefore the potential exists for any of the immune effector mechanisms described in Chapters 8 and 9 to be responsible for the actual tissue damage. Autoantibody bound to cells in the tissue can fix complement, resulting in direct complement-mediated damage. Complement activation can also stimulate an inflammatory response with infiltration into the tissue and activation of neutrophils and monocytes, causing further tissue damage. Antibody can also act as an opsonin, promoting phagocytosis, or contribute to ADCC. CD8 T cells can kill target cells presenting autoantigens on their class I MHC molecules. Finally CD4 T cells can act as effector cells in promoting a delayed-type hypersensitivity response leading to the recruitment and activation of monocytes that cause tissue damage not involving antibody.

The difficult task in these autoimmune diseases is sorting out the actual effector mechanism from the potential mechanisms. Unfortunately the presence of a particular cell type in an infiltrate, or the presence of autoantibody, does not necessarily mean it is contributing to the pathology. In experimental IDDM, autoantibodies against β cell antigens are produced but do not appear to contribute to the destruction of the β cells. This means that for many autoimmune diseases, including the examples above, the nature of the damaging autoimmune effector mechanism is not known. It is also quite possible that for many autoimmune diseases more than one effector mechanism contributes to the tissue damage.

12.4 Aetiology of autoimmune disease

The factors that contribute to the development of autoimmune disease are very complex. It has become clear in recent years that both genetic and environmental factors contribute to the development of autoimmune disease. Autoimmune diseases tend to run in families, so if one family member has an autoimmune disease there is an increased likelihood of another member having the disease. However, because families tend to live together, it is not always clear whether it is the sharing of genes or the environment that increases the incidence of an autoimmune disease in families.

Some of the best evidence for both genetic and environmental contributions to the development of autoimmune disease comes from the study of **concordance** between identical and non-identical twins. 'Concordance' is a term used to describe the incidence of a particular disease or trait in twins. This is easiest to understand if you consider a situation where a disease is caused solely by inherited genetic factors. In this situation, if one identical twin gets the disease the other twin will always get the disease because they are genetically identical. The concordance in this situation is 100%. Non-identical twins share on average 50% of their genes and therefore the concordance for a purely genetically determined disease would be 50% if one gene caused the disease and less if more than one gene was involved. Because twins are usually brought up in the same household and environment it is safe to assume that the difference in concordance in identical and non-identical twins is due to genetic factors.

If a disease is due to a combination of genetic and environmental factors the concordance for identical twins will be less than 100% but it will still be higher than for non-identical twins. This is the case with autoimmune diseases where concordance between identical twins is about 25% and that for non-identical twins is nearer 5%. The exact percentages vary from autoimmune disease to autoimmune disease but identical twins always show higher concordance than non-identical twins. These studies show that susceptibility to autoimmune disease can be inherited and therefore has a genetic element.

Concordance studies also show another very important feature of autoimmune diseases. As mentioned above, if a disease was 100% due to inherited genetic factors the concordance between identical twins would be 100%. The fact that concordance in identical twins is much less than 100% indicates that environmental factors are also contributing to the development of autoimmune diseases. Other evidence also points to environmental factors contributing to the development of autoimmune disease. The incidence of autoimmune disease has risen dramatically in the past 40–50 years, especially in North America, Europe and Australasia. Although some of this increase may be due to better diagnosis it is clear that there is a real increase and that this has happened too quickly to be explained by genetic changes in the population. Furthermore, populations from countries with a low incidence of certain autoimmune diseases, such as Japan and Papua New Guinea, have seen a rapid rise in the incidence of the autoimmune diseases when they have moved to countries with a higher incidence; again this cannot be explained by genetic factors and must be due to environmental effects.

12.4.1 Genetics of autoimmune disease

Genetic studies have shown that autoimmune diseases are multigenic. This means that many genes contribute to susceptibility to the disease. This is different from diseases, such as cystic fibrosis, which are caused by mutations in a single gene.

So what do we mean when we say a gene contributes towards susceptibility to a disease? Many genes are polymorphic; that, is they are present in the population in more than one form. Each form of a particular gene is called an allele. Polymorphism can occur in the part of the gene coding for the protein. This polymorphism can lead to different forms of the protein being made and is called **structural polymorphism**. The most prominent form of structural polymorphism is in the MHC genes where many different forms of the protein can be made (see Chapter 4). The second type of polymorphism occurs in the part of the gene not coding for the protein. This does not affect the structure of the protein but, if the polymorphism is in the promoter/enhancer part of the gene, it can alter the level of expression of the gene and affect how much of the protein is produced. This is called **non-structural polymorphism**. Polymorphism can therefore affect the biological activity of a protein in two ways: (i) by causing the production of different forms of the protein and (ii) by affecting the amount of protein produced (Figure 12.7). If it can be shown that people with one allele of a polymorphic gene have a different incidence of a disease than people with a different allele of the gene, the gene is clearly affecting susceptibility to the disease and is called a susceptibility gene. This is presumably due to the difference in amount or activity of the gene product.

Figure 12.7 Polymorphism affects the biological activity of the gene product. Structural polymorphism occurs in part of the gene coding for the protein and may result in slightly different forms of the protein, which may affect the biological activity of the protein, e.g. a hormone or its receptor. Non-structural polymorphism occurs in part of the gene that does not code for the protein so that, although the protein structure is not affected, the amount of protein produced may be different for the different alleles of gene, e.g. the polymorphism may affect the binding of a transcription factor.

The degree to which a particular allele of a gene increases susceptibility to a disease can be quantified by calculating what is known as the relative risk, or RR. The RR compares the frequency of a particular allele of a gene in a population with a disease with the frequency of the allele in the population as a whole. Although the calculation is more complicated the relative risk can be seen as the following ratio:

$$RR = \frac{\text{Frequency of allele in population with autoimmune disease}}{\text{Frequency of allele in population without autoimmune disease}}$$

Modern advances in DNA technology have made it possible to perform genome-wide screening for susceptibility genes in autoimmune diseases. These screens, combined with animal studies, show that in some autoimmune diseases up to 20 genes may contribute to susceptibility. These susceptibility genes can be divided into two categories – MHC and non-MHC.

MHC genes

It was realised many years ago that one of the strongest associations between genes and autoimmunity was with the MHC genes, especially class II MHC. This is perhaps not too surprising given that class II MHC proteins present antigen to CD4 T cells. Autoimmune responses are almost always CD4 T cell-dependent in the same way that nearly all specific immune responses against foreign antigens require CD4 T cells for their generation. The relative risk for some HLA alleles with autoimmune disease is shown in Table 12.2. It has been estimated that MHC genes contribute about 50% of the total genetic risk of developing autoimmunity.

Non-MHC genes

Although the most significant susceptibility gene for nearly all autoimmune diseases is MHC, many other genes also contribute to susceptibility to

Table 12.2 MHC association with autoimmune diseases

Disease	MHC association	Relative risk (RR)
Goodpasture's syndrome	DR 2	16
Multiple sclerosis	DR 2	5
SLE	DR 3	6
Myasthenia gravis	DR 3	3
RA	DR 4	7
Hashimoto's thyroiditis	DR 5	3
IDDM	DR 3	3
IDDM	DR 4	3
IDDM	DR 3 + 4	14

autoimmunity. One set of genes that affects susceptibility to autoimmunity is the **sex-related** genes. Many autoimmune diseases are much more frequent in women than men: Hashimoto's thyroiditis is 50 times more frequent in females and SLE occurs 10 times more frequently in women. Although the immunological basis for this is unknown, experimental studies have shown that it is in part due to the effect of sex hormones on the immune system.

Other genes are less well characterised. While new DNA technology has made it possible to identify the number of genes and their approximate chromosomal location, identifying the exact susceptibility gene is a time-consuming, arduous and expensive business. Some genes have been identified that contribute to autoimmunity in general and are associated with a number of autoimmune syndromes. People with mutations in the AIRE gene (see Section 11.5) suffer from a number of autoimmune syndromes including thyroiditis and type 1 diabetes. This is thought to be due to the fact that AIRE controls expression of organ-specific proteins by thymic medullary epithelial cells and people with a non-functioning AIRE protein have deficient expression of these genes in their thymus and therefore defective clonal deletion of T cells specific for these autoantigens. Another gene associated with generalised autoimmunity is the CTLA-4 gene. CTLA-4 is a protein that is found on activated CD4 T cells and can also exist in soluble form. CTLA-4 is a negative regulator of T cell function; it binds to CD80 and CD86, as does CD28 (see Section 7.2.2) but when CTLA-4 binds to CD80/86 it delivers a negative signal to the T cell and inhibits activation. People with defective CTLA-4 function are deficient in their ability to regulate autoreactive T cells and are at increased risk of autoimmune diseases such as thyroiditis, type 1 diabetes and autoimmune Addison's disease, which results in damage to the adrenal glands. In terms of non-MHC genes associated with specific autoimmune diseases, the only non-MHC gene accepted so far is the insulin gene in IDDM. This has a polymorphism in the non-coding part of the gene and different alleles of the insulin gene are associated with different susceptibility to IDDM. Several studies have shown that different alleles of the thyroglobulin gene may affect susceptibility to thyroid autoimmune disease; thyroglobulin is an important autoantigen in thyroid disease. Other genes suspected of being susceptibility genes fall into the category of cytokines and their receptor genes, apoptosis-related genes and genes coding for autoantigens. It is thought that these genes affect susceptibility to autoimmunity by affecting tolerance induction and the type of autoimmune response generated.

12.4.2 Environment and autoimmune disease

Again, while the evidence for environmental effects contributing to the development of autoimmune disease is substantial, it is much more difficult to identify what the environmental influences are. Part of the problem is the chronic nature of autoimmune diseases. An autoimmune response may be initiated by an environmental event and cause damage to tissue for a number of years before any clinical symptoms develop and the person sees a doctor. Therefore it will not be known exactly when the autoimmune response began and it will be difficult, if not impossible, to associate an environmental event with the triggering of an autoimmune response.

Furthermore, people may not remember too clearly events that happened to them years ago. However, some behavioural traits are associated with increased incidence of a particular autoimmune disease and allow some environmental candidates to be identified. The main types of environmental factor associated with autoimmunity include the following:

- **Infectious agents.** The best-known example of infectious agents contributing to autoimmunity is rheumatic fever, where antibodies against streptococcal M-antigen react against heart myosin, joints and kidney, resulting in arthritis and heart disease (see Section 12.5.1).

- **Drugs.** Drugs or their metabolites can bind to self-antigens and make them appear foreign. This can result in the development of autoantibody against the self-antigen itself (see Section 12.5.1). Examples of this are penicillin metabolites, which bind to red blood cells, and sedormid, which binds to platelets.

- **Toxins and pollutants.** In an autoimmune disease called Goodpasture's syndrome, autoantibodies are produced against type IV collagen, which is present in the basement membranes of the kidney and lung. All individuals with this disease suffer from glomerulonephritis, which leads to impaired kidney function. Cigarette smokers, but not non-smokers, also have a very high incidence of pulmonary haemorrhage, which can be fatal. This is thought to occur because the cigarette smoke damages the lungs, making the basement membrane accessible to the autoantibodies.

- **Food.** There have been a number of reports associating different foods (e.g. cows' milk and diabetes) with autoimmune diseases but none has been substantiated.

Because so few environmental agents have been identified it is difficult to know exactly how they contribute to the development of autoimmune disease. The general assumption is that they contribute to the loss of immunological tolerance that is the central feature of autoimmune diseases.

12.5 Loss of immunological tolerance

Autoimmune disease is caused by either the production of high-affinity autoantibodies or the generation of autoreactive T cells. Everyone has B cells with specificity for some self-antigens but they normally do not produce autoantibody because there are no self-antigen-specific Th cells (see Chapter 7). If autoreactive Th cells are generated because of a loss of T cell tolerance they can help autoreactive B cells to make antibody, and autoimmunity will follow. However, there are ways in which autoreactive B cells can make antibody in the absence of autoreactive Th cells, and these will be described first.

There are two ways in which B cells can make antibody to a self-antigen in the absence of Th cells specific for that self-antigen. One, which is quite simple to understand, is to bypass the need for Th cells. The other, which is more complicated, is due to cross-reactive epitopes between self-antigen and antigens on infectious agents. These processes occur as follows.

Th bypass by B cell mitogens

Bacterial products, such as lipopolysaccharide (LPS), can directly stimulate B cells to proliferate and differentiate into antibody-producing cells without the need for Th. This stimulation is not antigen-specific and the agents are called **mitogens**. Because any B cell can be stimulated, B cells specific for autoantigens as well as B cells specific for foreign antigens will be stimulated (Figure 12.8a). This stimulation is transient and once the bacteria have been eliminated the stimulation of autoreactive B cells and production of autoantibody will stop.

Cross-reactive epitopes

Some proteins on infectious agents have epitopes that are very similar to self-epitopes and this can lead to the production of autoantibody in the following way. Let us take a hypothetical self-antigen that has on it a self-B cell epitope and a self-T cell epitope (Figure 12.8b). B cells exist that are specific for the self-B cell epitope. They can bind the antigen through the B cell epitope, process the antigen and present the self-T cell epitope on their class II MHC. However, because there are no Th specific for the self-T cell epitope, the B cell will not receive the help it needs to make antibody (Figure 12.8b).

Now imagine an antigen on a pathogen that has the same B cell epitope as the self-antigen but has foreign T cell epitopes. This antigen can also be bound by self-reactive B cells; these B cells process the antigen and now present *foreign* antigenic peptide on their class II MHC (Figure 12.8b). Because the antigenic peptide is foreign there will be Th cells present that are specific for the foreign epitope and therefore the B cell will receive the T cell help and can be stimulated to produce autoantibody. The best known example of this is streptococcal M-antigen, which has B cell epitopes that are cross-reactive with cardiac myosin. Following infection with group A streptococci some individuals make antibodies against myosin, resulting in damage to the heart and rheumatic fever.

Another scenario is that a drug or environmental chemical can bind to a self-antigen, modifying it and creating a T cell epitope that appears foreign. Again the autoreactive B cell binds to the B cell epitope and presents the foreign-looking T cell epitope to Th cells and autoantibody is produced (Figure 12.8b).

In these situations, where pathogenic antigens or chemicals provide the foreign T cell epitopes, autoantibody production is usually transient. Once the pathogen has been eliminated or the person has ceased contact with the

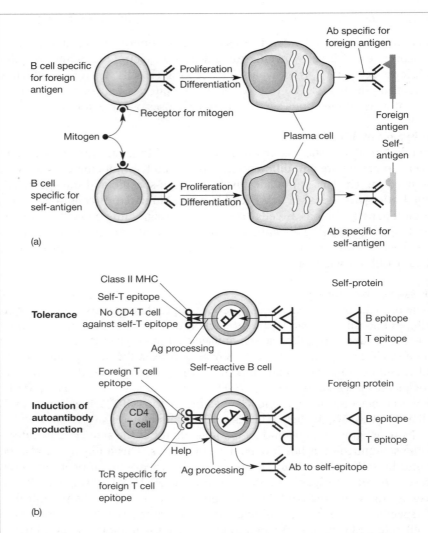

Figure 12.8 Induction of autoantibody production. (a) Autoantibody production induced by B cell mitogens. B cell mitogens stimulate B cells to proliferate and differentiate into plasma cells irrespective of the antigen specificity of the B cell, leading to production of autoantibody as well as antibody to foreign antigens. (b) Induction of autoantibody by cross-reactive epitopes. In the top example a self-protein is depicted with a self-B cell epitope and self-T cell epitope. Autoreactive B cells with specificity for the B cell epitope may bind the self-protein through the B cell epitope, take up the antigen, process it and present the self-T cell epitope on their class II MHC. However, because there are no CD4 T cells with specificity for the self-T cell epitope, the B cell receives no help and is not activated. In the bottom panel a foreign protein (e.g. from a pathogen) shares the same B cell epitope as the self-protein but has a foreign antigen T cell epitope. The same autoreactive B cell binds to the B cell epitope on the foreign protein and presents the foreign T cell epitope on its class II MHC. Because the T cell epitope is foreign there will be CD4 T cells with specificity for the T cell epitope and the B cell will be activated to produce antibody that reacts with the B cell epitope on both the foreign protein and the self-protein.

drug or chemical, the foreign T cell epitope is no longer present and the autoreactive B cell no longer gets T cell help. However, in genetically susceptible individuals, environmental triggers can lead to the activation of Th cells that are specific for a self-antigen; therefore the autoimmune disease becomes self-perpetuating even in the absence of the environmental agent. Why these autoreactive CD4 T cells are present and how they become activated is a subject of much investigation. At present the answer is unclear but a number of mechanisms have been suggested.

12.5.2 Activation of autoreactive Th

The emergence of pathogenic autoimmune Th cells is probably a multistage process that reflects a complex interplay between the many genes and environmental factors described in Section 12.4.

Genetic factors

Originally it was thought that loss of self-tolerance in CD4 T cells occurred because the normal mechanisms of preventing self-tolerance broke down owing to the presence of 'faulty' genes. The reason so many genes were involved in determining susceptibility to autoimmunity was because of the number of different tolerance mechanisms (deletion, anergy and regulation). All of them would have to stop working for autoreactive Th to be generated.

This view of the multigenic nature of autoimmune disease has now been modified, mainly because of a surprising set of observations (see Box 12.2 for more details). These studies suggest that there is heterogeneity in how different people maintain tolerance to the same self-antigen. For any given self-antigen some people may clonally delete CD4 T cells that are specific for the self-antigen but some will not. Although they do not automatically develop autoimmunity, people who do not delete self-reactive CD4 T cells are clearly at more risk of these cells causing an autoimmune response than people who do delete those particular self-reactive CD4 T cells.

The mechanisms that different people use to maintain tolerance to a particular autoantigen must be under genetic control and, although they are unknown, the genes controlling tolerance will affect susceptibility to autoimmunity. It is now thought that genetic susceptibility to autoimmunity is due to the inheritance of a specific combination of alleles of many different genes. In addition to controlling self-tolerance these genes may also control other factors associated with an immune response, such as cytokine production or apoptosis. The number of genes is not known but is quite large – 20 different genes contribute to susceptibility to IDDM or SLE. Because you have to have a specific combination of alleles of many genes to be at risk of developing an autoimmune disease, most people will have some susceptibility alleles but not enough to be at risk. Those people with the combination of enough alleles are at risk of getting an autoimmune disease; whether they get it or not depends on environmental factors.

BOX 12.2: VARIATION IN SELF-TOLERANCE

Dispersed throughout the genome are sequences of DNA called minisatellites, which are sequences of 10–100 bps that are repeated in tandem arrays of 0.5–40 kb. Some minisatellites are polymorphic in that the number of repeats varies. These are called **variable number tandem repeats**, or VNTRs. Because they affect the DNA structure, VNTRs near genes may affect expression of the gene product. There is one such VNTR 5' to the insulin gene and different alleles of this VNTR affect susceptibility to IDDM. How the different alleles affect susceptibility to IDDM is not known but one intriguing possibility is that they affect expression of insulin in the thymus. Insulin expression has been detected in the thymus of some, but not all, individuals at both mRNA and protein level. This surprising observation has been seen with other proteins that are specific for extra-thymic tissue. Different individuals express some, but not all, of these tissue-specific proteins in the thymus. The implication of these observations is that people who express a peripheral self-antigen in the thymus will clonally delete thymocytes that are specific for these self-antigens but others will not. This means that for many (all?) peripheral self-antigens you may have the situation that different people are relying on different forms of self-tolerance to the same auto-antigen. The different forms of tolerance carry different risks of auto-reactive cells being activated *in vivo*. People who clonally delete self-reactive CD4 T cells against a self-antigen clearly carry the smallest risk of generating Th against the antigen. However, people who have auto-reactive CD4 T cells in their periphery are at much greater risk of generating an autoimmune response against this autoantigen.

Why should there be such variety in the way people maintain self-tolerance? Surely the safest bet would be to evolve mechanisms to clonally delete all self-reactive CD4 T cells. The best explanation for this variation relates to the balance between the threat posed by pathogenic organisms and the threat posed by autoimmunity. Individuals who clonally delete CD4 T cells specific for a self-antigen may also delete cells specific for a similar antigen on a pathogen. These individuals have a reduced risk of autoimmunity but also a reduced response to the pathogen and therefore an increased risk of infection. Conversely individuals who have 'less vigorous' tolerance to a specific self-antigen have an increased risk of autoimmunity but would give a stronger response to the pathogen and therefore be more likely to survive infection. Within a population there will be a variety of potential responses to each autoantigen, and therefore to different pathogens, so that of the population as a whole a balance is achieved between the two threats.

Environmental activation of autoreactive Th

Although many experimental procedures can lead to the activation autoreactive Th in animals it is not clear exactly how environmental facto lead to the activation of autoreactive Th in human autoimmune disease One aspect that seems to be very important is the involvement of dendrit cells. Autoreactive CD4 T cells, like those against any antigen, need to initially activated by antigen presented to them on class II MHC express by dendritic cells (see Chapter 6). Two main mechanisms have been suggested to explain how environmental agents could promote the activation of autoreactive CD4 T cells by dendritic cells (Figure 12.9):

- **Molecular mimicry.** An environmental agent (such as an infectious organism) will be regarded as foreign by the immune system and dendritic cells will present antigenic peptides derived from the agent on their class II MHC. If one of these antigenic peptides is very similar to a self-antigen peptide, CD4 T cells specific for that peptide will be activated to become Th. These can then react against self-peptides as well as those from the environmental agent and cause autoimmunity.

- **Tissue damage.** Tissue damage, whether caused by a toxic substance or a pathogen, can lead to an inflammatory type of response with the production of cytokines and the release of self-antigens from damaged cells. These two factors can activate dendritic cells to present self-peptides derived from the self-antigens on their class II MHC, which could then activate self-reactive CD4 T cells.

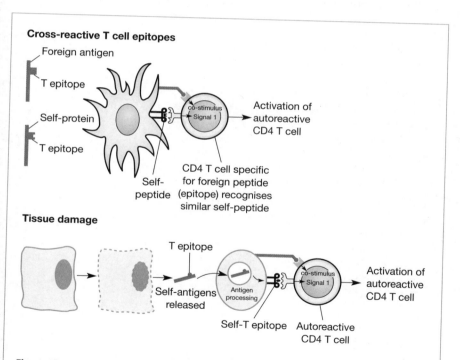

Figure 12.9 Activation of autoreactive Th cells. Activation of autoreactive CD4 T cells can occur due to a pathogen having a very similar epitope to a self so the CD4 T cells activated by the epitope on the pathogen react with the closely related self-epitope (top panel). Alternatively, tissue damage may result in the release of self-antigens, which are taken up and processed by dendritic cells and can stimulate autoreactive CD4 T cells.

12.6 Summary

- Autoimmune diseases constitute a large group of diseases that are characterised by an immune response against one or more components of the body. Overall they affect 5% of the population in industrialised countries.

- In some autoimmune diseases the pathology is solely due to the production of high-affinity autoantibody against self-antigens. In other diseases there is also an extensive inflammatory infiltrate, consisting of many different cell types in the affected tissue, making it difficult to identify the exact nature of the immune-mediated damage.

- The aetiology of autoimmune diseases is complex, with many genes and environmental factors contributing to susceptibility. One of the major genetic associations is with MHC class II. Few of the other susceptibility genes have been identified.

- The way in which the environment contributes to the development of autoimmune disease is not well understood but it is thought either to provide autoreactive B cells with a source of help for antibody production or to contribute to the activation of autoreactive CD4 T cells.

CHAPTER THIRTEEN

Allergy and other hypersensitivities

Learning objectives

To learn about the different ways in which inappropriate antibody responses can cause disease. To understand the immunological, clinical, genetic and environmental aspects of asthma and related allergies. To know about other types of hypersensitivities caused by antibody responses. To be aware of contact sensitivity reactions.

Key topics

- Type I hypersensitivity (allergy)
  ○ Immunological basis of type I allergic responses
  ○ Clinical symptoms of allergy
  ○ Genetics of allergy
  ○ Environmental factors
  ○ Treatment
- Type II hypersensitivity
- Type III hypersensitivity reactions
- Contact sensitivity

13.1 Introduction

Although antibody can provide vital protection from disease, there are situations where excessive production of antibody against harmless antigens actually causes disease. This is known as **hypersensitivity** and there are three types of antibody-mediated hypersensitivity. The most common and well known is type I hypersensitivity or allergy. This is mediated by the production of IgE and is commonly seen as an allergic response. Type II hypersensitivity is caused by cytotoxic antibodies against normal or modified tissue components and type III hypersensitivity is caused by the deposition of antibody–antigen complexes in blood vessels of various tissues. Neither type II nor type III hypersensitivity involves the production of IgE.

13.2 Type I hypersensitivity (allergy)

The most familiar syndrome associated with allergy is asthma. Approximately one in seven children in Britain suffers from asthma and the incidence has risen dramatically over the past 50 years. Until the twentieth century asthma was a rare condition. Like autoimmunity, asthma rates are highest in more industrialised societies.

Asthma is one clinical manifestation of diseases caused by the production of IgE against otherwise innocuous antigens and is an example of an allergic reaction. Antigens that stimulate an allergic response are called **allergens**. Allergens can be derived from several sources:

- **Pollens.** Many pollens cause allergy, especially rhinitis and asthma. Pollens from plant, trees and grasses can all cause allergy, with ragweed being a particular problem in North America.

- **Insects.** Allergens derived from the house-dust mite are the most common cause of asthma. Cockroaches are also a common source of allergen. Bee and wasp stings also stimulate strong allergic reactions, possibly because they contain pharmacologically active products.

- **Animals and birds.** Animal dander (fur, excretory products) and feathers are another major source of allergens. Cats, dogs and rodents are the most common source of animal allergens although this probably reflects the degree of exposure rather than any special feature of these animals.

- **Drugs.** Many drugs cause allergic reactions, with penicillin and sulphonamides being among the most common.

- **Food.** A number of foods cause allergies with sometimes severe effects. Peanuts, shellfish and milk products are the best known causes of food allergies.

Allergic reactions can cause other illnesses apart from asthma: rhinitis (hayfever), eczema (rashes), conjunctivitis and diarrhoea are all examples of allergic reactions. These allergic reactions are localised and usually reflect the site of exposure. Airborne allergens are more likely to cause asthma and hayfever, while food allergens tend to cause gastric symptoms, although this is not always the case, e.g. food allergens can cause skin reactions. In its severest form an allergic reaction can be systemic, leading to anaphylaxis, an extremely dangerous condition that can be fatal if not treated. The present concern about peanut allergy is due to a number of deaths caused by anaphylactic reactions to foods containing peanut products.

The use of the terms '**atopy**' and 'allergy' can be a source of confusion. Although often used synonymously, the two terms do have slightly different meanings. Allergy is described as an IgE-mediated hypersensitivity to a particular substance that manifests itself either clinically, in one or more of the forms described above, or as a positive skin prick test (see Section 13.4). Atopic individuals are those who have demonstrable evidence of IgE hyperreactivity. Although this would include allergic individuals, it would also include individuals who have specific IgE against an allergen but have no clinical symptoms and a negative skin prick test. Most atopic individuals are, however, allergic.

13.2.2 Immunological basis of allergy

The development of allergy occurs in a number of stages, the first of which is **sensitisation**.

Sensitisation

Sensitisation involves the production of IgE in response to exposure to allergen (see Figure 13.1). This IgE production is dependent on CD4 T cells that respond to antigens associated with the allergen and differentiate into Th2 cells. Secretion of IL-4 by the Th2 cells is particularly important in making B cells specific for the allergen switch to IgE. The IgE binds to mast cells, which have receptors called FcεRs on their surface that bind to the Fc portion of IgE. Mast cells are found in most tissues and there are two types, mucosal mast cells and connective tissue mast cells. Although they are quite similar, the two types of mast cell do differ in a number of respects (Box 13.1). IgE can stay bound to mast cells for months and possibly years. The initial exposure to allergen does not usually involve any symptoms.

Mast cell activation

Upon re-exposure to the allergen, the allergen will bind to the allergen-specific IgE that is bound to the FcεRs on mast cells. This will cause cross-linking of the FcεRs, resulting in a series of intracellular signalling events that lead to mast cell activation. The immediate consequence of mast cell activation is degranulation and the release of preformed mediators from the mast cell granules. This is followed by the synthesis and release of newly formed mediators (Figure 13.1). A list of preformed and newly formed mediators, together with their main functions, is given in Table 13.1. These mediators cause vasodilation, increased vascular permeability and smooth muscle contraction, as well as having other effects. The symptoms caused by these events depend on the anatomical site and are described below. These events generally occur within minutes of exposure to the allergen.

Figure 13.1 The allergic response. Upon first exposure to antigen, CD4 T cells specific for the antigen are stimulated to become Th2 cells ①, which help B cells differentiate into plasma cells secreting allergen-specific IgE ②. The IgE binds to FcεRs in mast cells and basophils ③. A second or subsequent exposure to the allergen results in the allergen binding to the IgE on the mast cells, which causes cross-linking of the FcεRs, resulting in mast cell activation ④. The immediate response of mast cell activation is degranulation of the mast cells with the release of the granule contents ⑤; these are preformed mediators such as histamine and heparin. The activated mast cells also begin to synthesise new mediators ⑥, especially leukotrienes, prostaglandins and cytokines (TNFα).

BOX 13.1: MUCOSAL AND CONNECTIVE TISSUE MAST CELLS

Although histologically they look very similar, mucosal mast cells (MMC) and connective tissue mast cells (CTMC) differ in their location, T cell dependency, expression of FcɛR, content of inflammatory mediators and sensitivity to drugs, as indicated in the table below.

Properties of mucosal and connective tissue mast cells

Property	Mucosal mast cells	Connective tissue mast cells
Location	Gut and lung	Most tissues
T cell dependence	Yes	No
Relative histamine content	Low	High
Relative FcɛR expression	High	Low
Major protease	Tryptase	Chymotryptase
Relative leukotriene C4 content	High	Low
Relative prostaglandin D2 content	Low	High
Inhibition by sodium cromoglycate	No	Yes

Some of these differences have only been detected in mice and it remains to be determined whether all the differences occur in humans.

Table 13.1 Inflammatory mediators released by mast cells

Mediator	Main biological effects
Preformed (released upon degranulation)	
Histamine	Vasodilation, increased vascular permeability
Enzymes (proteases)	Degradation of blood vessel endothelial membrane
Eosinophil chemotactic factor	Attract eosinophils from blood
Neutrophil chemotactic factor	Attract neutrophils from blood
Tumour necrosis factor-α	Promote inflammation
Newly synthesised after mast cell activation	
Prostaglandins	Vasodilation, platelet aggregation
Leukotrienes	Increased vascular permeability, mucus secretion
Platelet activating factor	Aggregation and degranulation of platelets
Tumour necrosis factor-α	Promote inflammation

Chemotactic factors attract different cell types

Accumulation of eosinophils, basophils, neutrophils and T cells

TNFα and other factors

Vasodilation + increased vascular permeability

Different adhesion molecules allow binding of eosinophils, basophils, neutrophils and T cells

Figure 13.2 Late phase reaction. This occurs a few hours after exposure to the allergen. TNFα secreted by mast cells activates local endothelial cells, which up-regulate expression of adhesion molecules that promote the migration of eosinophils, basophils, neutrophils and T cells to the site. These cells are activated to give a late phase inflammatory response.

Late phase reactions

Activated mast cells secrete TNFα, which activates the endothelium at the site causing expression of adhesion molecules that promote the migration of leucocytes from the blood (see Chapter 2). Chemotactic factors, including PGD2, LTB4, IL-8, MIP1α and eotaxin, are also produced. The net effect of these factors is the later recruitment of eosinophils, basophils, neutrophils and T cells to the site. Activation of these cells produces further inflammation, which is seen as a late phase reaction (Figure 13.2).

13.3 Clinical symptoms of allergy

Allergic responses can be localised to the site of allergen or spread throughout the body and become systemic. The clinical symptoms of localised allergic reactions depend on where the mast cells are localised and can result in asthma, hayfever, diarrhoea/vomiting or skin reactions (eczema and hives). A systemic allergic reaction is called anaphylaxis.

13.3.1 Local allergic reactions

Asthma
The main symptom of asthma is shortness of breath. This is caused by narrowing of the large and small airways, a condition called bronchospasm. The narrowing of the airways can be due to three factors (Figure 13.3 and

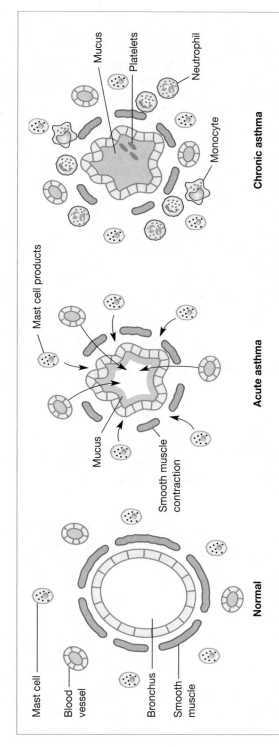

Figure 13.3 Asthma. Acute asthmatic attacks are due to the release of histamine, LTC4 and PGD2, which cause smooth muscle contraction of the lower airways (bronchi). Histamine and LTC4 also stimulate mucus secretion, which contributes to narrowing of the airways. The late phase asthmatic reaction is due to the influx of other inflammatory cells that stimulate further mucus secretion, epithelial cell damage and the activation of platelets leading to microthrombi in the airways.

Plate 16). Histamine, LTC4 and PGD2 released by mast cells cause smooth muscle contraction and therefore bronchoconstriction. Histamine and LTC4 also increase mucus secretion into the airways, which further narrows the available passage for air flow. The late phase reaction results in an inflammatory exudate consisting of eosinophils, neutrophils, basophils, macrophages and platelets, which causes further blockage of the airways.

Prolonged exposure to the allergen can lead to a condition known as bronchial hyper-reactivity. In this condition the bronchi are exquisitely sensitive to histamine and other mediators so that bronchospasm and asthmatic attacks can be triggered by stimuli such as exercise, cold air or passive cigarette smoke without exposure to the allergen. If the individual can avoid the allergen for a period of time this hyper-reactivity may disappear.

Rhinitis (hayfever)

Hayfever is characterised by a blocked and runny nose (rhinorrhoea), coughing, sneezing and itchy eyes and is the result of activation of mast cells in the nose and conjunctivae of the eye. Because the allergen is often pollen, hayfever can be seasonal and occur only during the time of year when a particular pollen is present. The mast cell mediators cause localised vasodilation and increased vascular permeability. This may be accompanied by the entry of mucosal mast cells into the nasal mucosa, where normally only connective tissue mast cells are found. The presence of both types of mast cells can complicate the chemotherapy of rhinitis since neither cell type responds to all anti-allergic drugs (see Section 13.7.2).

Food allergies

Activation of mast cells in the gut causes smooth muscle contraction, vasodilation, increased secretion of fluids and reduced peristalsis, resulting in vomiting and/or diarrhoea. The inflammation of the mucosa can also allow food allergens to enter the bloodstream, where they can activate mast cells at other sites, causing asthma or skin allergies.

Skin allergies

Activation of mast cells in the skin results in vasodilation and increased vascular permeability. These cause increased blood flow and oedema, which cause itchy red swellings (called hives or urticaria) that, if chronic, lead to a severe itchy skin rash (eczema).

13.3.2 Anaphylaxis

This is the result of a systemic allergic reaction and can be life-threatening. Depending on its severity it can cause symptoms similar to those seen in localised allergic reactions or it can be a life-threatening condition known as anaphylactic shock. The mildest manifestation of anaphylaxis is hives (urticaria). More severe forms result in swelling of the lips, tongue and larynx, nausea, vomiting and asthma. Anaphylactic shock causes severe swelling of the larynx (laryngeal oedema), severe asthma and/or severe hypotension due to massive loss of fluid from the blood to the tissues, which can lead to circulatory collapse. Death can result from respiratory or circulatory failure.

13.4 Testing for allergy

The most common test for allergy is the skin prick test. In this test a minute (a few microlitres) amount of allergen is introduced into the epidermis using a lancet. If a person is sensitised to the allergen they will give a **wheal and flare** reaction (Plate 16). The basis of the wheal and flare reaction is the same as that for eczema or hives. The allergen cross-links the FcεRs on skin mast cells causing mast cell activation. The mast cell mediators cause vasodilation, which increases blood flow to the area causing redness or a flare. The mediators also cause increased vascular permeability, which makes the blood vessels leaky. The leakage of blood fluid into the tissue results in oedema and swelling, the wheal. The mast cell mediators also cause itching. The flare is seen within seconds or minutes and the wheal occurs shortly after. The immediate wheal and flare reaction usually disappears in less than an hour after application of the allergen but can be followed by the late phase reaction (Section 13.2.2). This reaction typically occurs four to six hours later and is seen as a lump, which may be accompanied by pain (Plate 16). The main histological feature is the accumulation of eosinophils, basophils and neutrophils at the site.

Skin tests usually give good agreement with allergic status. However, some people with clinical allergy to known allergens fail to give a positive skin test to the allergen and others give a positive skin test to allergens but display no clinical signs of allergy to the same allergen.

Total and antigen-specific IgE can be measured by ELISA (Section 12.3). Antigen-specific IgE is now considered to be a more reliable predictor of atopy than total IgE.

13.5 Epidemiology of allergy

Allergy, like autoimmunity, has dramatically increased in incidence in the past 50 years. Particularly alarming has been the rise in the incidence, severity and mortality due to asthma, especially in children. Again, like autoimmunity, both genetic and environmental factors contribute to susceptibility to allergy. Family studies show that although the overall rate of atopy is 15%, the incidence rises to 30% in children with one atopic parent and 50% where both parents are atopic, suggesting a genetic component to susceptibility. Concordance in identical twins is about 60%, indicating that environmental factors are also involved.

13.5.1 Genetics of allergy

Another feature that allergy shares with autoimmunity is that it is multigenic, i.e. many genes determine susceptibility to developing allergy. Genes affect many aspects of the allergic response, some of which are immunological and some non-immunological. Not surprisingly for an immunologically based syndrome, both MHC genes and non-MHC genes are involved.

MHC genes

MHC genes appear to influence both overall IgE levels and the IgE response to individual allergens. An example of the latter is the response to allergens in rye grass where an increased IgE response is seen in people who are *HLA-DR3*. IgE responses to other allergens are associated with other *HLA-DR* alleles. The ways in which MHC genes affect overall IgE responses are not clearly understood.

Non-MHC genes

A number of candidate genes affecting susceptibility to allergy have been proposed. Some of these affect IgE production and some affect other aspects of the allergic response.

- **IL-4.** A polymorphism associated with the IL-4 gene has been linked with overall IgE levels. This is not surprising, given the importance of IL-4 in promoting B cells to switch to IgE. It should be pointed out that the IL-4 gene is located on chromosome 5q in a cluster containing genes for IL-3, IL-5, IL-9, IL-13 and granulocyte-monocyte colony stimulating factor, as well as IL-4. It cannot be excluded that the IgE association is not with one or more of these genes instead of, or in addition to, the IL-4 gene.

- **FcεR.** The FcεR consists of four chains, an α-chain, a β-chain and two identical γ-chains (Figure 13.4). The β-chain is located on chromosome 11q and genetic markers have revealed an association between allergy and the region of chromosome 11 containing the gene for the β-chain. However, it has not yet been proved that the linkage is actually with the β-chain.

- **Non-immunological genes.** Other genes may not be involved with the immune response to the allergen but may affect the physiological response to allergen-induced mediators. None of these genes has been definitively identified but one interesting candidate is the β2-adrenergic receptor, where polymorphisms associated with this gene could affect the responsiveness of smooth muscle and endothelial cells.

13.5.2 Environmental factors in allergy

Genetic studies, and the sudden dramatic increase in the incidence of allergy, demonstrate that environmental factors contribute to the development of allergy. Environmental factors could affect both the incidence and the manifestations of allergy. However, although environmental factors are important, it is not clear how they have contributed to the sudden alarming increase in allergy incidence.

Three main environmental factors have been suggested as contributing to the increase in allergy. Modern lifestyles may have increased exposure to

Figure 13.4 The high-affinity receptor for IgE (FcεR). The α-chain of the FcεR binds to the Fc portion of IgE. The disulphide-linked homodimer of γ-chains is responsible for initiating intracellular signalling following cross-linking of FcεRs by allergen and is linked to the α–chain by the β-chain, which spans the cell membrane four times.

allergens; this may be especially true for the house-dust mite, a major cause of allergies. Changes in housing, with central heating, fitted carpets and better sealing of houses, have provided an ideal breeding ground for house-dust mites. A second possibility is increasing pollution, although clear-cut associations have not been shown between any single pollutant and allergy. Finally it has been suggested that modern vaccination strategies have changed the pattern of childhood infections and this has contributed to the increase in allergy. These possibilities are discussed in more detail in Box 13.2.

13.6 Why have IgE?

With all the problems associated with IgE production to innocuous substances, it would seem that IgE causes more problems than it solves. However, IgE would presumably not have evolved if it was not of some benefit, and certainly not if it was more harmful than beneficial. The situation where IgE production is seen most prominently, apart from allergy, is in parasitic worm infections. However, it has to be said that the evidence that IgE is beneficial in worm expulsion is not extensive. One of the consequences of triggering of mast cells by IgE is the recruitment of eosinophils to the site. Eosinophils are able to kill worms *in vitro* by antibody-dependent cell-mediated cytotoxicity (Section 8.8) and it is assumed this can also occur *in vivo*. The extensive activation of mucosal mast cells and eosinophils during an IgE-mediated inflammatory response could also make the gut an unhealthy place to be from a worm's point of view, possibly leading to parasite expulsion. Therefore IgE may have evolved to deal with parasitic worm infections, which affect billions of people and animals worldwide. In evolutionary terms the immune system has not learnt to deal with the rapid change in lifestyle experienced in the past 100 years or so, so that in industrial societies, where

BOX 13.2: CONTRIBUTION OF THE ENVIRONMENT TO ALLERGY

A number of environmental factors have been proposed as contributing to the increase in allergy.

Presence of the allergen

Although it may seem obvious, one reason for the increase in allergy is increased exposure to allergens. One of the most common allergens is the house-dust mite and a number of studies have shown a link between exposure to house-dust mites and the development of atopy.

Factors affecting IgE production

A number of environmental factors have been implicated in promoting IgE production. These include the following:

- **Air pollutants.** Diesel exhaust particles demonstrate the difficulty in identifying environmental factors in allergy. In experiments on mice, diesel exhaust particles have been demonstrated to act as adjuvants for IgE production. However, there is generally not a good correlation between allergy and environmental exposure to diesel fumes. This does not mean that diesel exhaust particles are not playing a role in the human disease but that other factors complicate the analysis.

  Other air pollutants implicated in the development of allergy are ozone, nitrogen dioxide and tobacco smoke, although they may be bigger factors in triggering asthmatic attacks rather than the development of allergy.

- **Other pollutants.** Mercuric chloride, present in some drug preparations, cosmetics and disinfectants, directly stimulates Th2 cells. Since these produce IL-4, which promotes IgE production, a mechanism for promotion of allergy by mercuric chloride can be seen although there is no direct evidence that this occurs in humans. Platinum and manganese salts have also been suggested as contributing factors to allergy development.

- **Changes in infection patterns.** One interesting idea that has been put forward to explain the recent rise in asthma, seen especially in industrial societies, is that, owing to changes in living environments and the practice of increased vaccination, children are now exposed to a different spectrum of infection. Until 40 years or so ago most children were exposed to many viral diseases such as mumps, chickenpox, measles and rubella (German measles) and many were exposed to polio or smallpox. These are intracellular infections that would be expected to stimulate Th1 responses. Exposure of children to allergens in this Th1 environment may result in a Th1 response to the allergen and therefore not favour the Th2 response that promotes IgE production. Nowadays, not only are children vaccinated against these diseases but the only adjuvant used in humans is alum, which promotes a Th2 response. Therefore children are exposed to allergen in the relative absence of Th1 responses and in the presence of increased Th2 responses, which could promote the production of IgE in response to the allergen.

parasitic infections in humans are less prevalent, the balance of IgE production has swung away from benefit to harm.

13.7 Treatment of allergy

Treatment of allergy can be divided into three main approaches: allergen avoidance, pharmacological treatment and immunological intervention.

13.7.1 Allergen avoidance

A number of studies have shown that if individuals can avoid contact with allergens their clinical symptoms improve and can disappear. However, it is not always possible to identify the allergen and, if identified, to avoid it totally. Allergen avoidance does not eliminate the underlying state of allergy, so that exposure to the allergen, even after many years of avoidance, can lead to reappearance of the symptoms.

13.7.2 Pharmacological intervention

Pharmacological treatment of allergy is targeted at preventing mast cell degranulation or inhibiting the inflammatory response associated with allergic reactions. These agents treat the symptoms of allergy but do not change the underlying immune response and therefore do not provide a cure.

The most common treatments for allergy are antihistamines, which block the actions of histamine, and inhibitors of mast cell degranulation. Drugs such as sodium cromoglycate and nedocromil inhibit mast cell degranulation but it is not clear whether they work on all mast cell types.

More severe allergies are treated with steroids. These may be given locally, i.e. applied to the skin or inhaled, or systemically. They act primarily by inhibiting the production of cytokines or other inflammmatory mediators by various cell types, including eosinophils and monocytes. They may also inhibit the migration of inflammatory cells into the site of the allergic reaction.

Finally adrenaline is given to treat anaphylactic shock. These agents bind to β-adrenergic receptors on smooth muscle cells and cause smooth muscle relaxation, leading to a reversal of the vasodilation and increased vascular permeability seen during anaphylaxis. It cannot be used as a long-term treatment for allergy.

13.7.3 Immunological intervention

In contrast to pharmacological approaches, immunological intervention attempts to modify the immune response against the allergen so that the production or effects of IgE are negated. A number of different approaches are now being used or developed, some of which are described below.

Desensitisation (hyposensitisation) involves giving small amounts of allergen to the patient to induce the production of IgG against the allergen.

If the individual subsequently encounters the allergen, the IgG binds to the allergen and stops the allergen binding to IgE on mast cells, thereby preventing mast cell activation (Figure 13.5). Although it can be effective, the procedure has the major drawback that giving too much allergen may trigger a potentially fatal anaphylactic attack.

Another approach is to give anti-IgE antibodies to block IgE from binding to the FcεR on mast cells and activating them. For reasons that are not understood, it was found that the anti-IgE antibodies inhibited the production of IgE as well as inhibiting its activity; this may be due to anti-IgE binding to IgE on B cells and inhibiting them or causing their opsonisation. This approach has been successful at reducing allergic responses to the house-dust mite in clinical trials.

13.8 Type II hypersensitivity

Type II hypersensitivity involves the production of IgG or IgM antibodies that react with antigens on cells or tissues. There are three major

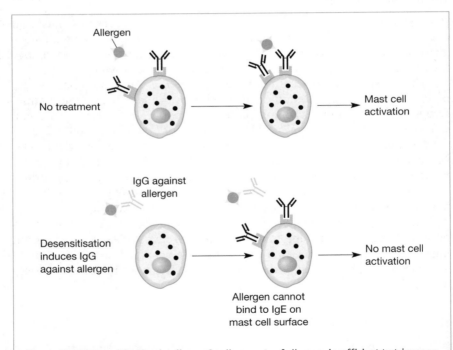

Figure 13.5 Desensitisation in allergy. Small amounts of allergen, insufficient to trigger an allergic response, are given repeatedly to an allergic individual to stimulate production of IgG specific for the allergen. If the person is accidentally exposed to a large amount of allergen (enough normally to produce an allergic response) the IgG binds to the allergen and prevents the allergen binding to the allergen-specific IgE on the mast cells, thus preventing mast cell activation and an allergic attack.

syndromes associated with type II hypersensitivity: these are blood transfusion reactions, haemolytic disease of the newborn and drug-induced hypersensitivity. Some autoimmune diseases involve the production of autoantibodies against cells or tissues, causing a type II hypersensitivity-like reaction (see Chapter 12).

13.8.1 Blood transfusion reactions

It was appreciated early in the twentieth century that it was not possible to transfuse blood at random between individuals. There are over 20 antigens on erythrocytes, and they vary between individuals. The most important and well known are those that comprise the ABO blood group system. Differences in glycosylation of red cell glycoproteins give rise to three variants of the glycoprotein called A, B or O. The groups are co-expressed so that individuals can be O, A, B or AB (Figure 13.6). These carbohydrate groups are similar to those seen on bacteria and therefore even if someone has not been exposed to a foreign blood group through blood transfusion, they may develop antibodies against the bacterial antigens that also recognise foreign blood groups. Individuals with blood group O make antibodies to groups A and B; individuals with blood group A make antibodies to B; and those with blood group B make antibodies to A. These antibodies are usually of the IgM class. AB individuals do not make antibodies to A or B and can receive blood of any type.

If someone receives an incompatible blood group to which they have antibodies, the antibodies bind to the transfused red blood cells. IgM is very good at fixing complement and complement fixation results in activation of the classical complement pathway and lysis of the transfused red blood

Figure 13.6 ABO blood groups. Differential glycosylation at the end of red blood cell glycoproteins result in blood group O, which contains the core glycoprotein only, blood group A, which has a terminal *N*-acetylgalactosamine, or blood group B, which has a terminal galactose.

cells. This can result in the release of massive amounts of haemoglobin, some of which may be metabolised to bilirubin, which is toxic at high levels. Clinical symptoms include fever, chills, nausea and vomiting, and chest and lower back pains.

Blood transfusion reactions can also be caused by antibodies to minor blood group antigens. However, these antibodies do not usually arise through cross-reactivity with microbial antigens but instead are the result of repeated transfusions.

13.8.2 Haemolytic disease of the newborn

Another antigen on red blood cells is called the rhesus (Rh) antigen. Most individuals are Rh+ but a minority lack the rhesus antigen and are Rh–. During childbirth considerable amounts of foetal blood enters the mother's bloodstream. If an Rh– mother has an Rh+ baby the woman will be immunised against the Rh+ antigen during childbirth and produce IgG antibodies to Rh (Figure 13.7). If the mother has another Rh+ baby, the anti-Rh IgG can cross the placenta and bind to the foetus's red blood cells. This can result in opsonisation of the red blood cells and their phagocytosis and destruction in the liver or spleen. This results in an enlarged spleen and liver and toxicity due to bilirubin. In severe cases the condition can be fatal.

Sensitisation of the mother can be prevented by giving the mother a preparation of antibodies against the Rh antigen shortly after birth. These antibodies bind to the foetal red blood cells and cause their immediate removal and destruction before they have time to sensitise the mother.

13.8.3 Drug-induced hypersensitivities

Some drugs or their metabolites, can bind to red blood cells or platelets. Although the drugs are too small themselves to stimulate an immune response, by binding to self-proteins the drug–self-protein complex can create a new antigen that appears foreign to the immune system. In a small proportion of people these new antigens stimulate antibody production against the drug. The antibodies bind to the drug and cause either complement-mediated lysis or opsonisation through Fc receptors and C3 receptors on phagocytes (Figure 13.8). Some drugs, such as penicillin, bind to red blood cells resulting in anaemia; quinidine can bind to platelets, resulting in thrombocytopenia. Usually, when the drug is withdrawn the symptoms resolve as non-drug-bound blood cells replace the drug-bound ones.

13.9 Type III hypersensitivity

Type III hypersensitivity reactions are caused by immune complexes of antigen and antibody. They are seen in a number of situations and can be local or systemic (Figure 13.9).

①

②

③

Anti-Rh Ab
in maternal
blood

Placenta

rbc

Maternal
Abs bind
to rbcs

Maternal
circulation

Placenta

Placenta

Maternal
circulation

| Rh– mother
Rh+ baby | Childbirth → | rbcs from baby
enter maternal blood,
stimulate αRh-IgG | 2nd pregnancy → | αRh-IgG crosses
placenta and causes
lysis or opsonisation
of foetal
rbcs → anaemia |

Figure 13.7 Haemolytic disease of the newborn. If a mother who is rhesus-negative (Rh−) has a rhesus-positive baby (Rh+), there is the possibility that at birth some of the baby's red blood cells (rbcs) will enter the mother's circulation ② and stimulate the production of antibodies to the Rh antigen. If the mother has another Rh+ baby, the maternal IgG antibodies against Rh will cross the placenta and enter the baby's blood circulation where the antibodies will bind to the Rh antigen on the baby's rbcs, resulting in complement activation and lysis of the rbcs, causing anaemia.

Systemic immune complex disease

The historical example of systemic immune complex disease is serum sickness. Animal serum (usually horse), containing antitoxin antibodies, was given as a treatment for diphtheria or tetanus to neutralise the toxins secreted by the bacteria. Because the serum came from another species, the serum proteins stimulated a strong antibody response. If the person was given a second dose of serum, the antibodies bound to the serum proteins forming immune complexes. These complexes are deposited in blood vessel walls, especially those of the kidney, skin and joints, where they fix complement, leading to an inflammatory response. The resulting symptoms include fever, rashes, arthritis and kidney malfunction. Although serum sickness can now be avoided, systemic immune complex disease can occur in some chronic infectious diseases, such as leprosy, malaria, hepatitis and streptococcal infection.

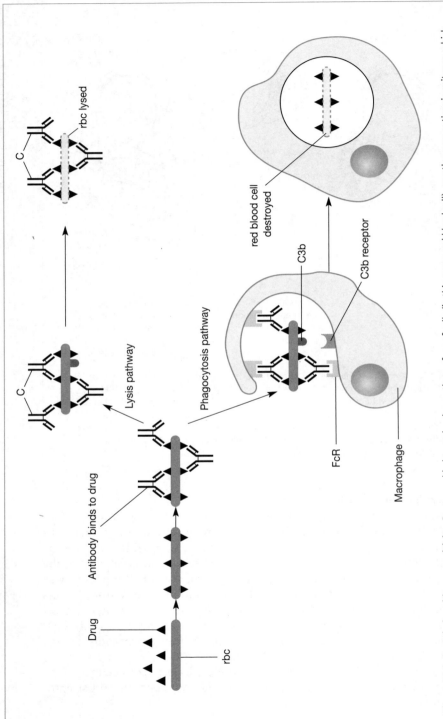

Figure 13.8 Drug-induced hypersensitivity. Drugs can bind to molecules on the surface of cells (in this case a red blood cell), creating new antigenic epitopes, which stimulate antibody formation. The antibody binds to the drug on the cell surface, resulting in complement activation and lysis or opsonisation of the red blood cell.

Figure 13.9 Type III hypersensitivity. Ab–Ag complexes are deposited on the wall of small blood vessels ① where they activate complement ②. C3a and C5a activate basophils ③ which release vasoactive amines (e.g. histamine) ④. The vasoactive amines cause vasodilation and increased vascular permeability ⑤, which exposes the basement membrane, allowing platelet binding and activation and the formation of microthrombi ⑥. The C3a and C5a activate neutrophils ⑦, which secrete degradative enzymes causing damage to the basement membrane. If the response persists, the inflammatory mediators enter the tissue where they activate mast cells, initiating an inflammatory response ⑧.

Localised immune complex disease

Type III hypersensitivity can also cause a localised disease. This is mos commonly seen after repeated inhalation of antigen. The two best-know examples are farmer's lung, caused by the inhalation of mould spores o hay, and pigeon fancier's disease, caused by the inhalation of pigeon anti gens in dried faecal particles.

13.10 Differences between type II and type III hypersensitivity

There is often confusion about the differences between type II and type II hypersensitivity, which both involve the same Ig classes (usually IgG) an complement activation leading to inflammation. The crucial differenc between type II and type III hypersensitivity is the nature of the antigen. I type II hypersensitivity the antigen is expressed on the target tissue that i affected by the disease, e.g. in haemolytic disease of the newborn the rhesu antigen is on the red blood cells that are removed or destroyed by the anti body. By contrast, in type III hypersensitivity the antigen is not associate with the target tissue. Type III hypersensitivity is the result of the formatio of immune complexes of antibody and antigen. It is not the nature of th antigen that determines where type III hypersensivity reactions occur. Th sites of type III hypersensitivity reactions are determined by where th immune complexes are deposited, and this is usually in small blood vesse such as those seen in the joint, skin and kidneys.

13.11 Contact hypersensitivity

Contact hypersensitivity reactions (CS) occur in response to encounter (small chemicals with the skin. This results in an eczematous reaction invol ing swelling, redness and severe itching, which occurs 24–72 hours afte contact with the agent in a sensitive individual and is known clinically as con **tact dermatitis**. One of the best-known causes of CS reactions, especially i the USA, is poison ivy. The leaves of the poison ivy plant contain pentadec catecol, a small chemical that stimulates a very strong CS reaction. Othe causes of CS reactions are some metals, such as nickel, gold and chromiun used in jewellery, and chemicals found in rubber. The actual chemicals tha stimulate CS reactions are small, often with a molecular mass of less than 10 DA. These chemicals are too small to stimulate an immune response in the free form and are called **haptens**. These haptens have the ability to chemicall attach themselves to proteins of the host where they modify the proteins, cr ating new antigenic epitopes that stimulate the CS response.

CS reactions are very similar in nature to delayed-type hypersensitivi reactions (see Section 9.3) in that they involve primarily T cells and mono cytes but not antibody. However, whereas delayed-type hypersensitivit reactions play a protective role against some intracellular pathogens, C reactions usually occur against harmless agents and as such are detrimental to the host. Furthermore there seems to be more involvement of CD8 T cells in CS reactions, although their role is not clear. The CS reaction is initiated by haptenated-host proteins being taken up by Langerhans cells in the skin. The Langerhans cells migrate to the draining lymph nodes, where they stimulate hapten-specific CD4 T cells to divide and differentiate into Th and also stimulate hapten-specific CD8 T cells. A week or so after the initial encounter the sensitised CD4 T cells migrate to the site of contact and enter the tissue where they are stimulated by hapten being presented on class II MHC by dendritic cells and tissue macrophages. The Th are stimulated to release cytokines (IFNγ, TNFα, TNFβ, IL-3 and GM-CSF) and chemokines that attract monocytes to the site, resulting in swelling and redness and some tissue damage due to macrophage activation. CD8 T cells also enter the site, although it is not clear whether they contribute to the damage by directly killing hapten-bearing host cells or by secreting cytokines. Upon subsequent encounter with the hapten the reaction is more severe because of the pre-existing sensitised hapten-specific T cells. The response can then involve severe blistering because the extensive activation of macrophages (and possibly involvement of CD8 T cells) results in more damage and separation of skin cells.

13.12 Summary

- There are three types of antibody-mediated hypersensitivities: type I, type II and type III.

- Type I hypersensitivity involves the production of IgE against usually innocuous antigens such as pollen, house-dust mite faeces or food.

- The IgE binds to Fc receptors for IgE (FcεRs) on the surface of mast cells. Upon re-exposure to the allergen, allergen cross-links the FcεRs, leading to mast cell degranulation.

- Depending on the site of mast cell degranulation, allergy can be manifested by asthma, hayfever, vomiting and diarrhoea, hives, eczema or anaphylactic shock.

- Both genetic and environmental factors contribute to the development of allergy. The incidence has increased dramatically in the past 50 years.

- Treatment of allergy involves allergy avoidance, antihistamines, inhibitors of mast cell degranulation, steroids and immunological intervention, such as desensitisation or administration of anti-IgE antibodies.

- Type II hypersensitivities are caused by IgG or IgM antibodies reacting against antigens on cells or tissues, causing opsonisation and complement fixation.

- There are three major types of type II hypersensitivity reactions: blood transfusion reactions, haemolytic disease of the newborn and drug-induced hypersensitivities.

- Type III hypersensitivity reactions are caused by the formation of immune complexes, which are deposited and initiate inflammatory reactions. Antigens may be benign, such as foreign serum proteins (causing serum sickness) and pigeon faecal antigens (causing pigeon fancier's lung), or because of chronic infection with pathogens, such as malaria, leprosy or hepatitis.

- A fourth type of hypersensitivity reaction is contact hypersensitivity, which does not involve antibody but involves CD4 and CD8 T cells and monocyte/macrophages.

CHAPTER FOURTEEN

AIDS

Learning objectives

To know about the history, incidence and clinical progression of AIDS. To understand the biology of the AIDS virus and how it causes immunosuppression. To learn about the immune response to the AIDS virus and be aware of forms of chemotherapy and issues of vaccine development in AIDS.

Key topics

- History of AIDS
- The human immunodeficiency virus (HIV)
  ○ Structure
  ○ Replication cycle
- Clinical course of HIV infection
- Immunology of HIV infection
  ○ Effects of HIV on the immune system
  ○ Immune response to HIV
- Chemotherapy of AIDS
- HIV vaccines

14.1 History and incidence of AIDS

The story of AIDS began in 1981 when a cluster of unusual diseases was observed in certain groups of people. The two main diseases were pneumonia caused by a yeast, *Pneumocystis carinii*, and an unusual tumour called Kaposi's sarcoma. These diseases were seen initially in homosexual men but later the same symptoms appeared in intravenous drug users and haemophiliacs who were injecting blood-clotting factors. This pattern of disease occurrence suggested that a transmittable agent was responsible for the diseases. The unusual aspect of these diseases was that they are usually only seen in immunosuppressed people and not in people with a fully functioning immune system. The observation that individuals with these diseases had low numbers of CD4 T cells was consistent with immunosuppression and in 1982 the term **acquired immunodeficiency syndrome**, or AIDS, was used by the Centers for Disease Control in Atlanta to describe the disease.

In 1983 the virus that causes AIDS was isolated from the lymph node of an infected individual by Montagnier's group in Paris and was called the **human immunodeficiency virus**, or HIV for short. A second strain of HIV was identified in 1986; this was called HIV-2 and the first strain was renamed HIV-1. HIV-1 and HIV-2 differ in their virulence and geographical location. HIV-2 is less virulent than HIV-1 and is found primarily in western Africa. Genetic studies have shown that both HIV-1 and HIV-2 are natural viruses of primates that have jumped species to infect humans. HIV-1 came from chimpanzees and HIV-2 from the sooty mangabey. Both of these animals are killed for food and it is assumed that it was during this process that the virus initially infected humans. HIV-1 and HIV-2 do not cause immunosuppression in chimpanzees or sooty mangabeys; only when the virus crossed into humans did it cause the profound immunosuppression seen in AIDS.

14.1.1 Current incidence of HIV infection and AIDS

It is estimated that by the end of 2005 over 65 million people worldwide were infected with HIV and over 24 million will have died. The highest rate of increase in HIV infection is seen in sub-Saharan Africa, where an estimated 20–40% of young adults are infected. As a result of this, the life expectancy in sub-Saharan Africa has almost halved; it is now in the 30s–40s instead of approaching 70, which would have been the estimated life expectancy if the AIDS pandemic had not occurred. The pattern of spread in Africa appears to be primarily by heterosexual contact and has a similar incidence in men and women. By contrast, spread in Europe, the USA and Oceania is still mostly among 'high-risk' groups such as homosexuals and intravenous drug users and is much more prevalent in men than women. However, there are signs in the USA that the pattern of spread is changing, with more women being infected.

14.2 The human immunodeficiency virus

14.2.1 Strains and variants

Some of the terminology used to describe HIV can be confusing. The two strains of HIV are called HIV-1 and HIV-2. There are also many different subtypes of HIV-1 so vaccines will be needed to protect against all subtypes. Additionally, HIV has a very high mutation rate, giving rise to different forms of the virus known as variants. These variants are important because, as described below, they differ in which cell types they can infect.

Although there are different strains, subtypes and variants of HIV, they are very similar in structure and replication and therefore will be described together and referred to collectively as HIV.

14.2.2 Structure of HIV

HIV is a retrovirus (Figure 14.1) and contains RNA as its genetic material. The HIV genome contains two molecules of single-stranded RNA, each bound by a molecule of reverse transcriptase. Within the genome are also a p10 protease and a p32 integrase. The genome is surrounded by a nucleo-capsid consisting of an inner layer of protein called p24 and an outer layer of protein called p17. The outer portion of the virus consists of a lipid enve-lope derived from the host cell membrane containing two viral proteins gp120 and gp41, which collectively are called viral **envelope proteins**.

14.2.3 Replication cycle of HIV

Like any other virus, HIV must infect a host cell before replicating with the viral progeny, leaving the cell to infect others.

Infection

HIV infects CD4+ cells. In addition to CD4 T cells, monocytes and den-dritic cells also express CD4, although at lower levels than T cells. There are two stages to viral infection: binding to the host cell, and fusion with the cell membrane to allow the virus to enter the cell (Figure 14.2). The

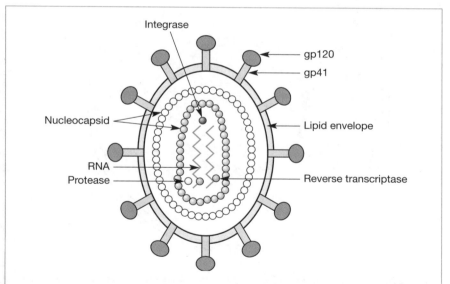

Figure 14.1 Diagram of an HIV particle. HIV is a retrovirus and has two single-stranded RNA molecules as its genetic material. The RNA is associated with reverse transcriptase, integrase and polymerase enzymes, which are necessary for viral DNA and RNA synthesis. Surrounding this is the nucleocapsid, which consists of an inner layer of p24 protein and an outer layer of p17 protein. The outer portion of the virus consists of a lipid layer derived from the host cell into which is inserted the viral gp41 envelope protein. Each gp41 protein molecule is associated with a molecule of the gp120 envelope protein.

initial binding of HIV to the host cell involves the gp120 protein on the surface of the HIV particle binding to CD4 on the host cell surface.

However, binding of g120 to CD4 is not enough for HIV to infect the cell. The gp41 component of the envelope protein binds to a second protein on the cell surface. This second protein differs between variants of HIV. In some variants the gp41 binds to the chemokine receptor CCR-5, which is on the surface of CD4 T cells, monocytes and dendritic cells. These variants can therefore infect all of these cell types and have been called M-tropic. Other HIV variants bind to another chemokine receptor called CXCR-4, which is present on CD4 T cells but not on monocytes or dendritic cells. These variants can only infect T cells and are called T-tropic. The CCR-5

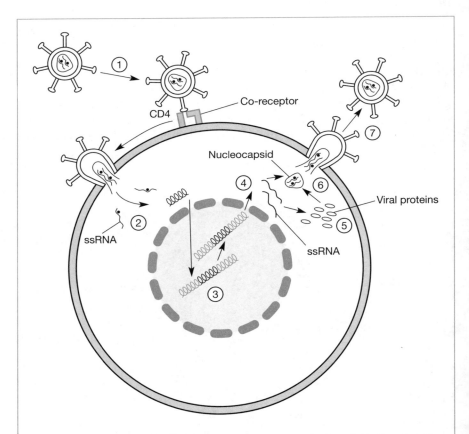

Figure 14.2 Life cycle of HIV. ① The virus binds to CD4 +ve cells through binding of gp120 to CD4 and interactions between the virus and chemokine co-receptors. ② The nucleocapsid enters the cell, where it unfolds, releasing viral RNA, which is reverse-transcribed to double-stranded DNA. ③ The viral DNA integrates in the host genome, where it lies dormant as a provirus. ④ Following cell activation, viral DNA directs the transcription of viral RNA. ⑤ Viral proteins are translated from the RNA. ⑥ The viral proteins and single-stranded viral RNA assemble to form new viral particles. ⑦ The virus buds from the cell, picking up some of the cell membrane, and the complete viral particles can infect other cells.

gene has two alleles, one of which results in a non-functional form of the CCR-5 protein. Approximately 1% of Caucasians are homozygous for the null allele and therefore express no functional CCR-5 protein. These individuals are resistant to infection with HIV. Since their expression of CXCR-4 is normal, this suggests that initial infection is by M-tropic variants using the CCR-5R. Indeed isolates of HIV in recently infected individuals are predominantly M-tropic. The emergence of T-tropic variants is often a sign of rapid progression to AIDS.

Following binding of Gp120 to CD4 and the relevant chemokine receptor, gp41 mediates fusion between the viral envelope and host cell membrane, allowing the viral nucleocapsid to enter the cell.

Replication

Once inside the cell the nucleocapsid of the virus is removed and the reverse transcriptase copies the RNA into double-stranded DNA. The DNA integrates into the host cell DNA, where it is known as a provirus. This stage of viral infection is known as the **latent** stage and the virus can lie dormant for a long time.

When the infected cell is activated, viral RNA is transcribed from the proviral DNA and viral proteins are translated using host cell protein synthesis machinery. The viral proteins and RNA assemble into particles that bud from the cell, taking some of the host cell membrane to form the viral envelope. These particles can now infect other cells.

14.3 Clinical course of HIV infection

The clinical progression of HIV infection can be divided into three stages: infection, the latent stage and development of AIDS (Figure 14.3).

14.3.1 Infection

Most people show no symptoms immediately after infection but about 15% demonstrate symptoms that are reminiscent of influenza: these include fever, malaise, aching muscles, sore throat and swollen lymph nodes. Some people also develop swollen lymph nodes without any other clinical symptoms. Following infection, antibodies to HIV antigens are produced, a process called **seroconversion**. The detection of antibodies to HIV is used to test for infection.

14.3.2 Latency

This period is generally asymptomatic although about 33% of infected people will have swollen lymph nodes. The average time from infection to the development of AIDS is about 10 years but the length of the latency period is extremely variable, lasting less than a year in some people and upwards of 15 years in others. It is not clear whether everyone infected with HIV will inevitably develop AIDS.

Figure 14.3 The development of AIDS. Following infection the generation of anti-HIV antibody and Tcs clears most of the virus and the disease enters the latent phase. After a variable period of time the loss of CD4 T cell function results in failure to control the virus and the development of AIDS.

14.3.3 Development of AIDS

The end of the latency period is accompanied by the emergence of various symptoms that indicate progression to AIDS without treatment. These symptoms include weight loss, night sweats, fever and diarrhoea. There are also infections such as oral candidiasis, herpes simplex and shingles caused by minor opportunistic infectious agents.

AIDS is defined clinically by the appearance of major opportunistic infections or by a drop in the CD4 T cell count to below 200 cells/µl of blood. The opportunistic infections are caused by all categories of pathogens as shown in Table 14.1. Without treatment AIDS invariably leads to death. Death is caused by combinations of the infections described in Table 14.1 although some infections are more prominent according to geographical location. In Europe, the USA and Oceania the commonest infection is pneumonia caused by *Pneumocystis carinii*. In Africa and Asia,

Pathogen	Type of organism	Disease
Candida albicans	Fungus	Thrush, disseminated mucocandidiasis
Pneumocystis carinii	Fungus	Pneumonia
Cryptococcus neoformans	Fungus	Meningitis
Herpes simplex	Virus	Pneumonia
Varicella-zoster (chicken pox)	Virus	Shingles, pneumonia
Cytomegalovirus	Virus	Pneumonia
Mycobacterium tuberculosis	Bacterium	Pneumonia
Salmonella spp.	Bacterium	Diarrhoea, septicaemia
Cryptosporidia	Protozoan	Diarrhoea
Toxoplasma gondii	Protozoan	Encephalitis

infections such as *Cryptosporidium*, a protozoan that causes severe diarrhoea and weight loss, and *Mycobacterium tuberculosis* are more common.

14.4 Immunological events associated with HIV infection

HIV has one of the most complicated relationships with the immune system of any infectious agent. Because of the nature of the cells it infects and the molecules that it binds to, HIV has a dramatic effect on the immune system, eventually causing profound immunosuppression. Before that stage, and contrary to earlier beliefs, the immune system, far from failing to mount a significant response against the virus, generates a very powerful response against HIV. With almost any other virus this response would be enough to eliminate the pathogen but because of the special features of HIV the virus is able to survive and eventually destroy the immune system.

14.4.1 Changes in the immune system during HIV infection

The hallmark of HIV infection is the gradual decrease in CD4 T cell numbers. However, there are a number of other changes to the immune system that occur following HIV infection.

Changes in lymph nodes
Many HIV+ individuals develop swollen lymph nodes during initial infection and this can persist even during the clinically latent phase. Histological examination of the lymph nodes shows increasing disruption of the normal lymph node architecture, an influx of CD8 T cells and eventual loss of germinal centres.

Loss of CD4 T cell function

As the infection progresses HIV+ individuals show a loss of CD4 T cell function that cannot be explained simply on the basis of reduced CD4 T cell numbers and is somehow associated with the effects of the virus (see Figure 14.3). Transplantation recipients taking immunosuppressive drugs can show drops in CD4 T cell numbers similar to those seen in HIV+ individuals but the transplant recipients demonstrate much better CD4 T cell function. The reasons for the loss of CD4 T cell function are not clear. One consequence of the loss of Th activity is the reduced ability to mount a delayed-type hypersensitivity reaction (Chapter 9) and as the infection progresses the ability to generate antibody responses is also lost.

Antibody abnormalities

Somewhat paradoxically, HIV+ individuals can demonstrate increased total levels of serum Ig despite the impaired ability to generate specific antibody responses. Again the basis for this is not clear but it is thought to reflect abnormalities in the normal regulation of immune responses caused by the virus. Possibly also related to generalised immune dysfunction, HIV+ individuals can show increased production of autoantibodies to antigens such as red blood cells, spermatozoa or myelin (a component of nerve sheaths) and they may suffer from flare-ups of allergies such as eczema.

14.4.2 Immune response to HIV

Immediately following infection with HIV there is a period of rapid viral replication resulting in high levels of virus in the blood (viraemia) (Figure 14.3). HIV is an intracellular pathogen and as such would be expected to stimulate a strong CD8 cytotoxic T cell response and possibly a delayed-type hypersensitivity response, as well as the production of antibody. In fact strong antibody responses to gp120 envelope protein and the p24 protein of the nucleocapsid (also known as gag) are seen soon after viral infection and there is also a powerful CD8 cytotoxic T cell against gp120, p24 and some of the proteins that make up the reverse transcriptase (pol antigens). The combined antibody and cytotoxic T cell response is extremely effective and eliminates more than 99% of the virus. Following this strong response the disease usually enters the latent stage.

It was initially thought that the latent stage of the disease was one in which the virus was largely in the proviral stage and therefore inactive and that very few CD4 T cells were being destroyed as a consequence of being infected by the virus. These conclusions were based primarily on studies examining the number of productively infected (meaning that whole virus could be cultured from the cells) CD4 T cells and overall CD4 T cell numbers in the blood of HIV-infected individuals. A number of recent studies suggest that this picture obtained from blood was not accurate.

Newer techniques to measure viral RNA, a sign that the virus is replicating, have shown that there is a persistent level of active viral replication in the blood even during the latent period. The level of this viraemia is a good prognostic indicator of the progression of the disease: individuals with a low level of viraemia progress more slowly to AIDS than those with a higher level of viraemia.

Examination of lymph nodes have shown 10–100 times more productively infected CD4 T cells than seen in blood. Furthermore, lymph nodes of HIV+ people in the latent period show structural abnormalities that become more severe as the disease progresses. Also during HIV infection the relative proportion of the total number of CD4 T cells in the blood increases so that the actual loss of CD4 T cells is greater than that reflected in the blood.

Treatment of infected individuals with drugs that prevented the virus from infecting new cells showed that (i) the immune system was eliminating up to 30% of the total viral load *every day* but that in the absence of these drugs the virus replicated at an enormous rate and the eliminated virus was replaced, and (ii) up to 2×10^9 CD4 T cells were being destroyed every day but most of these were replaced by the immune system so that the overall CD4 levels in the blood declined only very slowly.

It now appears that the latent stage of HIV infection may be clinically latent but actually represents a very dynamic situation with high viral destruction and replication and extensive CD4 T cell destruction and replacement. Two questions occur as a consequence of this picture: why does the immune response not clear the virus, and what is the cause of the extensive CD4 T cell destruction?

In terms of why the immune system does not clear the virus, HIV has a number of features that enable it to survive in the face of a powerful immunological onslaught:

- It has a very high replication rate. This is the main reason why the virus is not totally cleared.

- It can hide as a provirus where it is not detectable by the immune system.

- It has a very high mutation rate, so that antigens to which antibody and CD8 Tcs have been made mutate and are no longer recognised by the immune system. Antibodies and CD8 Tcs are made against the mutated antigen but this can mutate again and so the process of evading the immune response goes on and on.

The cause of CD4 T cell loss is thought to be a combination of direct killing by the virus and destruction of virally infected cells by the immune system (Figure 14.4). CD4 cells that are infected by the virus express viral antigens on their surface. These may be viral peptides presented by class I MHC or they may be soluble gp120 bound to CD4 on the T cells. HIV+ individuals can have large amounts of soluble gp120 in their blood and lymph, which will bind to CD4. Because of the expression of viral antigen, CD4 T cells can be killed in a number of ways (see Figure 14.4):

- **Antibody + complement.** Anti-gp120 binds to the gp120 bound to CD4, resulting in complement fixation and activation and cell lysis.

- **Antibody-dependent cell-mediated cytotoxicity.** Macrophages and NK cells have Fc receptors on their surface. This enables them to bind to the Fc portion of the anti-gp120 antibody bound to the CD4 T cell and kill the cell.

- **CD8 cytotoxic T cells** kill virally infected cells expressing HIV antigenic peptides in association with their class I MHC.

By generating these responses the immune system is actually doing what it is supposed to; it is generating effector mechanisms that eliminate virally infected cells and thereby the virus. Unfortunately the main type of infected cell is the CD4 T cell and therefore the virus kills the very cells needed to generate an immune response against it. Although the immune system makes a valiant effort to replace the CD4 T cells that are destroyed every day, the production of CD4 T cells never quite matches the destruction and there is a gradual drop in CD4 T cell numbers. Eventually the CD4 T cell

Figure 14.4 Killing of CD4 T cells. (a) HIV may directly cause lysis of CD4 T cells. (b), (c) Infected cells bearing HIV antigens may be killed by antibody and complement (b) or killed by antibody-dependent cell-mediated cytotoxicity (c). (d) HIV-infected cells presenting HIV peptides on their class I MHC molecules may be killed by CD8 Tcs.

numbers become too low to maintain normal immune responsiveness and immunosuppression ensues.

14.5 Chemotherapy of HIV

The high mutation rate of HIV is not only a problem for the immune system. It is also a problem in trying to treat HIV+ individuals with antiviral drugs. The first anti-HIV drug was zidovudine (AZT), a reverse transcriptase inhibitor (Figure 14.5) introduced in 1987. Although it inhibited replication of HIV, the virus very quickly mutated and became resistant to the drug.

There are now three types of anti-HIV drug in use clinically and combinations of two, or preferably three, drugs are of longer clinical benefit because there is less chance of the virus mutating to become resistant to three drugs at once. The three main categories of drug are nucleoside analogue reverse transcriptase inhibitors, non-nucleoside reverse transcriptase

Figure 14.5 Anti-HIV drugs. (a) Reverse transcriptase inhibitors interfere with the synthesis of viral cDNA from the viral RNA. (b) Protease inhibitors inhibit the cleavage of viral polyproteins into viral proteins. Without the viral proteins replication cannot occur.

inhibitors and HIV protease inhibitors (Figure 14.5). The two types of reverse transcriptase inhibitor act in different ways to inhibit the reverse transcriptase and therefore complement each other's action. The protease inhibitors inhibit HIV protease, an enzyme used to cleave a number of viral precursor polyproteins to produce many HIV proteins and enzymes (Figure 14.5). Without protease activity the virus cannot replicate.

Usually combination chemotherapy consists of two reverse transcriptase inhibitors and a protease inhibitor. This can be very effective at lowering levels of virus and raising CD4 T cell levels and results in significant clinical improvement so that many AIDS patients are able to leave the AIDS clinic and return home. The new therapy is also thought to be responsible for the sharp drop in death from AIDS seen in the USA since 1996. However, there are problems with combination therapy. There is considerable toxicity, especially to the bone marrow and gut, which means some people cannot take the drugs. The regime for taking the drugs is complicated and intrusive to a normal lifestyle. Some of the drugs have to be taken with food and some without, and some drugs cannot be taken within a certain time period of each other. Finally the drug treatment is very expensive – about $15,000 a year – so that it is not readily available in the countries that need it the most.

14.6 HIV vaccines

Attempts to generate vaccines for HIV have fluctuated in intensity over the years for a variety of different reasons, both economic and scientific. Initially two type of vaccines were considered:

- **Prophylactic vaccines.** The aim of these vaccines was to provide protection to individuals who were not infected with the virus; in this respect the aim was the same as for vaccines against other infectious agents such as measles, mumps, polio, smallpox, etc.

- **Therapeutic vaccines.** Because AIDS initially was considered a disease of immunosuppression where HIV-infected people did not make a very good response against the virus, it was though it might be possible to boost the immune response of HIV-infected people so that they could mount a response and clear the virus. With the current appreciation that most HIV-infected people make a very strong response against the virus it is not so clear how effective this approach might be.

The development of prophylactic vaccines against HIV has considerable difficulties of both an immunological and a logistical nature.

The main immunological issues concerning development of an HIV vaccine are which of the many different current approaches to vaccine development (see Chapter 10) to use, given the many subtypes and variants and the mutation rate of the virus, and what type of immune response to try and stimulate. Additionally there are no good animal models of HIV.

The main logistical problems are in testing the vaccines. Traditionally vaccines were developed against acute diseases that often occurred in waves or epidemics (e.g. measles or mumps) and had a low mortality rate. It was therefore comparatively straightforward to vaccinate a section of the population and see whether these people were protected from the disease compared with non-vaccinated individuals. It was possible this way to get an answer to the effectiveness of the vaccine in a relatively short time. Because AIDS is a chronically progressive disease the conventional approach would take many years to give an answer – too long for most people to wait. There are a number of vaccines in clinical trials at present using DNA, recombinant and antigen/peptide approaches, although none has demonstrated convincing protection so far.

14.7 Summary

- AIDS as a disease was first identified in 1981 in groups of high-risk individuals: homosexual men, intravenous drug users and haemophiliacs. Currently it is estimated that 30 million people are infected with the AIDS virus.

- The virus causing AIDS was isolated in 1983 and called the human immunodeficiency virus (HIV). A second strain, called HIV-2, was identified in 1986.

- HIV is a retrovirus, i.e. its genetic material consists of RNA. It infects CD4+ cells, mainly CD4 T cells but also monocytes and dendritic cells.

- Clinically HIV infection is characterised by the development of opportunistic infections, especially with the yeast *Pneumocystis carinii*, and the development of a rare tumour, Kaposi's sarcoma.

- The development of AIDS can take over 15 years from infection. Following infection and seroconversion there is a latent period that is largely aclinical.

- Although the latent period is aclinical there is an extensive amount of viral destruction and replication and also destruction and replacement of CD4 T cells. However, there is a gradual net loss of CD4 T cells, leading eventually to immunosuppression and the development of opportunistic infections, which are fatal without treatment.

- The most successful chemotherapy for HIV is triple therapy with a combination of reverse transcriptase inhibitors and viral protease inhibitors. However, this therapy has significant side-effects, is difficult to comply with and is very expensive.

- HIV vaccines are faced with the problem of the high mutation rate of the virus and logistical problems of testing the efficacy of a vaccine against a chronic infection.

CHAPTER FIFTEEN

Manipulating the immune system: transplantation and tumours

Learning objectives

To understand contemporary approaches to manipulating the immune system in terms of transplantation and immunotherapy of tumours.

Key topics

- Transplantation
  - Types of transplant
  - Transplantation antigens
  - Immune responses to transplants
  - Preventing graft rejection
- Tumour immunology
  - Tumour antigens
  - Antibody-based tumour therapy
  - Tumour vaccines

15.1 Introduction

Part of the reason for understanding as much as possible about how the immune system functions is to aid in the development of treatments that can manipulate the immune system to prevent or treat diseases in which the immune system is involved. One method of manipulating the immune system, that of vaccination against infectious disease (described in Chapter 10), has prevented millions of deaths and has removed the need for billions of children to go through the misery and danger of many common childhood illnesses such as measles, mumps, polio and diphtheria. However, manipulation of the immune system does not stop at vaccination. In situations such as transplantation and cancer it is also desirable to alter the way the immune system behaves, either to prevent graft rejection or to harness the power of the immune system to attack tumour cells. This chapter will cover some aspects of immune manipulation in the fields of transplantation and cancer.

15.2 Transplantation

Transplantation involves the transfer of tissue, cells or organs from one anatomical site to another. These two anatomical sites may be different sites on the same person, e.g. with a skin transplant for burns, or they may be equivalent anatomical sites in different people, e.g. a heart transplant from one individual to another. Transplantation of a variety of tissues, cells and organs is now an accepted medical procedure for treating a wide variety of diseases and further treatments using transplants are being developed (see Table 15.1).

15.2.1 Categories of transplant

As with so many other areas of immunology, transplantation has its own jargon. Terms have been used to describe the relationship between donor and recipient in transplantation:

- **Autograft.** A transplant of tissue from one site to another in the same individual is called an autograft. The most common instance is the transplantation of healthy skin to the burn area of burn victims to try to prevent infection and dehydration and promote wound healing.

Table 15.1 Types of clinical transplant

Transplant	Disease treated
Kidney	Nephritis, diabetic complications
Heart	Heart failure
Lung/heart + lung	Cystic fibrosis
Pancreas/islets of Langerhans	Insulin-dependent diabetes mellitus
Liver	Congenital defects of the liver, hepatitis, cirrhosis
Cornea	Cataracts
Skin (usually autologous)	Burns
Bone marrow	Congenital haematopoietic defects (e.g. thalassaemia, severe combined immunodeficiency), leukaemia, lymphoma

- **Isograft.** An isograft is a transplant performed between genetically identical individuals, e.g. identical twins. Like autografts, isografts will normally survive without any immunosuppression.

- **Allograft.** Transplants between non-genetically identical individuals of the same species are called allografts. This is the most common category of transplant and the graft will be rejected in the absence of immunosuppressive therapy.

- **Xenograft.** This involves transplanting tissue from one species to another, e.g. baboon to human or pig to human. As might be expected, xenografts are subject to the strongest rejection responses and pose extra problems in preventing their rejection.

15.2.2 Transplantation antigens

Transplants between non-identical donors and recipients (allografts or xenografts) are rejected because the immune system recognises foreign antigens on the graft and does what it is supposed to: mount a specific immune response to eliminate the foreign organ. The antigens that stimulate graft rejection are called, unsurprisingly, transplantation antigens.

The transplantation antigens that stimulate rejection of allografts fall into two categories: major transplantation antigens and minor transplantation antigens.

Major transplantation antigens
The major transplantation antigens are the class I and class II MHC molecules, HLA-A, B and C and HLA-DP, DQ and DR. As described in Chapter 4, MHC genes are highly polymorphic so that the donor and recipient will almost certainly have different alleles of at least some of the MHC loci. The cells of the graft will therefore have MHC molecules on their surface that are foreign to the recipient. Foreign (allogeneic) MHC antigens stimulate a particularly vigorous immune response (Figure 15.1). The reasons for this are not totally clear. Class I MHC alloantigens stimulate strong antibody and CD8 cytotoxic T cell responses. Class II MHC alloantigens stimulate CD4 T cells to become effector Th and also stimulate antibody responses. The Th help in the generation of antibody against class I and class II MHC, the generation of CD8 cyotoxic T cells, and the development of delayed-type hypersensitivity reactions (Figure 15.1).

Minor transplantation antigens
Proteins other than MHC are polymorphic, although to a much lesser degree than class I and II MHC. The graft donor and recipient will have different forms of many of these proteins which can act as antigens. Because they do not stimulate as strong a rejection response as MHC they are called minor histocompatibility antigens. Minor histocompatibility antigens are recognised by the recipient's T cells in association with self-MHC in the same way as any antigen from a pathogen (Figure 15.2). The self-MHC can be on the recipient's

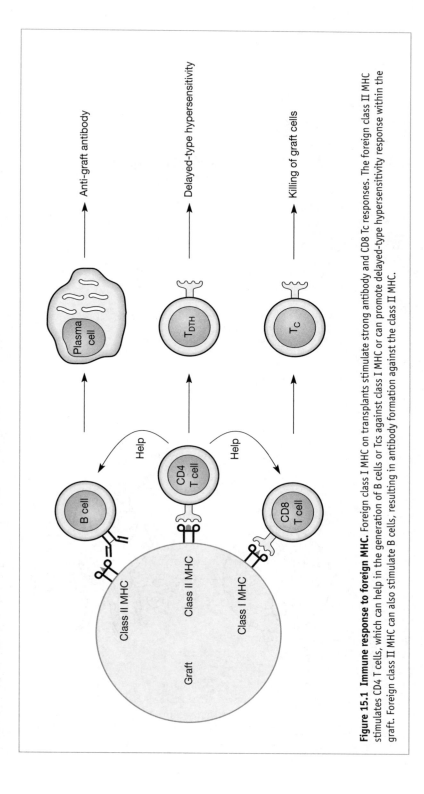

Figure 15.1 Immune response to foreign MHC. Foreign class I MHC on transplants stimulate strong antibody and CD8 Tc responses. The foreign class II MHC stimulates CD4 T cells, which can help in the generation of B cells or Tcs against class I MHC or can promote delayed-type hypersensitivity response within the graft. Foreign class II MHC can also stimulate B cells, resulting in antibody formation against the class II MHC.

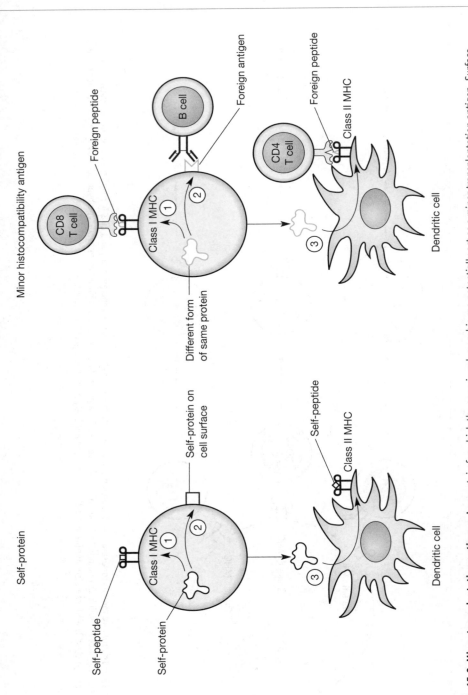

Figure 15.2 Minor transplantation antigens. Any protein for which the gene is polymorphic can potentially act as a minor transplantation antigen. Surface proteins can stimulate antibody production and internal proteins can be presented by class I MHC to simulate CD8 T cell responses. Protein that is taken up by dendritic cells can be processed and presented on class II MHC to stimulate CD4 T cells, which provide help for the antibody and CD8 T cell responses.

cells or, if the graft and recipient are matched for any MHC molecules, on the graft cells – in this case self- and graft MHC are shared. There are many minor transplantation antigens (probably more than 70), although few have been identified. Most stimulate only T cells although antibody is also made to antigens other than MHC on a graft.

15.2.3 The immune response against transplants

Graft rejection responses have been categorised into three main types: hyperacute, acute and chronic.

- **Hyperacute rejection** occurs if the recipient has pre-existing antibody against the graft. The antibody bind to antigens on the graft, resulting in complement fixation and activation of the clotting pathways. The graft essentially turns into a massive blood clot within minutes or hours of transplantation. Hyperacute rejection should be prevented by cross-matching (see Section 15.2.4).

- **Acute rejection** is the equivalent of a primary immune response against the graft. All kinds of effector mechanism can be generated against the graft: antibody, cytotoxic CD8 T cells and delayed-type hypersensitivity responses. Which of these occurs depends on a number of factors, including the degree of matching. They act against the graft in the same way as they act against pathogens, as described in Chapters 8 and 9. Without immunosuppression the graft would be rejected in 7–20 days.

- **Chronic rejection** is a process that usually starts after the graft has been in place for more than a year. One of the main pathological changes seen is thickening of the graft's arterial walls, which can lead to occlusion of the artery and ischaemia. Eventually a large enough part of the graft is without an adequate blood supply and the graft stops functioning.

15.2.4 Preventing graft rejection

There are three ways in which steps are taken to prevent graft rejection (Figure 15.3). First of all the donor and recipient are blood typed so that they are compatible for blood groups. They are also cross-matched to check that the recipient does not have pre-existing antibody to the donor. Cross-matching involves taking some serum from the recipient and adding it to leucocytes from the donor together with some complement. If the recipient has anti-donor antibodies they will bind to the donor leucocytes and fix complement, resulting in lysis of the leucocytes.

The second step to improving successful transplantation is matching the recipient and donor for HLA by tissue typing. HLA typing was described in Chapter 4. Unless the donor is a family member there will be almost no chance of matching the donor and recipient at all six HLA loci, especially taking into account that MHC molecules are co-expressed and therefore

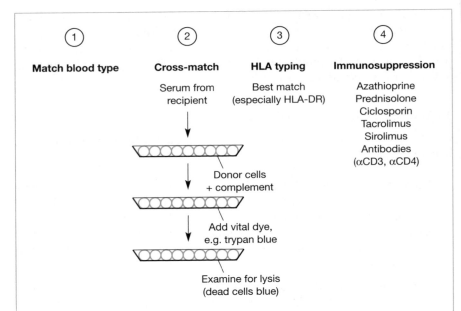

Figure 15.3 Steps in preventing transplant rejection. To prevent hyper-acute rejection caused by pre-existing antibody against donor antigens, the recipient and donor are matched for blood type and also cross-matched. Cross-matching involves incubating serum from the recipient with white blood cells from the prospective donor in the presence of complement. If the recipient has antibodies against donor antigens, the antibodies will bind to the donor cells and fix complement, resulting in lysis of the cells. The lysis can be detected by adding a vital dye such as trypan blue, which stains dead cells blue, while live cells are seen as white under a microscope. Potential donors and recipient are HLA-typed (see Chapter 4) and the donor with the best match is chosen for the transplant. Finally immunosuppressive agents are given to the recipient to try to prevent the anti-graft immune response.

people may well express two different alleles at one or more of the MHC loci. However, the more HLA matches there are between donor and recipient, the better the chance of a successful transplant. Matching at *HLA-DR* (Figure 15.4) has been shown to have the most beneficial effect but matching at *HLA-A* and *HLA-B* is also of benefit.

The third step in preventing graft rejection, and the one that made clinical transplantation feasible, is the use of immunosuppression. Antigen-specific immunosuppression is the suppression of the immune response to the graft without affecting the ability to respond to other antigens, including pathogens. This is still at the experimental stage with respect to transplantation. Non-antigen-specific immunosuppression involves the use of drugs or antibodies that inhibit immune responses against any antigen. Therefore, although the response to the graft is inhibited, so is the response to pathogens. This is the type of immunosuppression used clinically and transplant recipients are consequently at increased risk of infection and also of the development of some types of tumours.

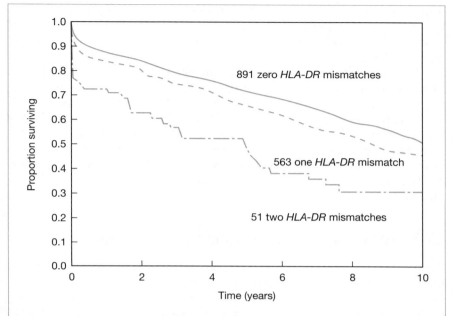

Figure 15.4 The effect of _HLA-DR_ matching on graft survival. Transplants where donor and recipient are matched at both _HLA-DR_ alleles show better survival than those with only one or no matches. The data are on kidney transplants performed in the transplant centre of Manchester Royal Infirmary from 1985 to 1999 (courtesy of Dr Phil Dyer).

15.2.5 Clinical immunosuppression

The first class of immunosuppressive drugs were the anti-mitotic agents. They were able to inhibit the proliferation of lymphocytes required during a specific immune response. The most commonly used drug in transplantation is azathioprine. One problem is that other tissues containing rapidly dividing cells, such as bone marrow and gut epithelium, are also affected.

Corticosteroids affect the transcription of many genes in lymphocytes and other cells. A consequence of this is that steroids affect lymphocyte migration, resulting in fewer cells entering the graft. They also inhibit macrophage and monocyte function and have general anti-inflammatory effects. The most commonly used steroid in transplantation is prednisone.

Up until the 1980s the immunosuppressive regimen of choice was a combination of azathioprine and prednisone. This could be quite effective and in some centres one-year kidney graft survival rates exceeded 80%. In the 1980s the first of a new breed of immunosuppressive drugs was introduced into clinical transplantation with dramatic effects on graft survival rates. This was **Ciclosporin**, a fungal product, which like the later members, tacrolimus and sirolimus, specifically inhibits CD4 T cell function. The drugs differ in the molecular details of their action but a major effect they have in common is to inhibit the **IL-2 pathway**. Ciclosporin and tacrolimus inhibit the production of IL-2, and sirolimus inhibits signalling through the IL-2 receptor. Because IL-2 plays a vital role in the proliferation of CD4 T cells, inhibition of the pathway provides a powerful immunosuppressive effect.

Finally antibodies can be used in transplantation to inhibit T cell function. These include polyclonal antithymocyte globulins from the serum of sheep or goats immunised with human thymocytes, or monoclonal antibodies (see Box 15.1) to T cell antigens such as CD3 or CD4. These antibodies act by causing the destruction of T cells or interfering with their function. They cause profound immunosuppression and therefore are used primarily in the period immediately after transplantation and to treat acute rejection episodes where the conventional immunosuppression is failing to prevent the rejection response.

15.3 Using the immune system against tumours

Tumour immunology has had a chequered past. In the nineteenth century it was noticed that some individuals with cancer showed a temporary remission (i.e. improvement) in the clinical status of their tumour during bacterial infection. Treatments were practised that involved deliberate infection of cancer victims with bacteria, a procedure that probably stood more chance of killing the patients than helping them. In the second half of the twentieth century the genetics of cancer became better understood and it was realised that cancer involved mutation of normal genes and that these mutated genes could give rise to altered proteins that could act as new antigens that were foreign to the individual (Figure 15.5). The concept of immune surveillance was born. This idea was that the immune system constantly patrolled the body looking for cancerous cells that expressed mutant antigens, which would be recognised as foreign. An immune response would then be generated against the mutant antigen leading to the destruction of the tumour cells. This led to a series of naive approaches in which it was thought that simply boosting the immune system would somehow increase the response against tumour cells and 'cure' cancer. These attempts were unsuccessful and brought the field of tumour immunology into a modicum of disrepute.

Today much more is known both about the immune system and the biology of cancer. This has led to a renewed interest in harnessing the immune response to provide protection against cancer. There are two main approaches that utilise the immune system to try to destroy tumour cells. These are the use of antibodies against tumour cells and the development of tumour vaccines.

15.3.1 Antibodies can be used to kill tumour cells

The approach of using antibodies against tumours is to make monoclonal antibodies (Box 15.1) that are specific for tumour antigens and to inject

Figure 15.5 Tumour antigens. Mutation of a gene in a tumour cell can lead to a different form of the protein, which is recognised as foreign by the immune system. Surface proteins can stimulate antibody production and internal proteins can be presented by class I MHC to stimulate CD8 T cell responses. Protein that is taken up by dendritic cells can be processed and presented on class II MHC to stimulate CD4 T cells, which provide help for the antibody and CD8 T cell responses.

BOX 15.1: MONOCLONAL ANTIBODIES

The old-fashioned way to make antibodies was to immunise an animal, usually a rabbit or sheep, with antigen and take its serum as a source of antibody. The problem with this approach was that the serum contained many different antibodies, some against the antigen, and some against pathogenic antigens that the animal had recently been exposed to. In the 1970s a technique was developed that enabled the production of large amounts of antibody derived from a single B cell. This involved taking B cells from an immunised animal and fusing the cells with immortalised myeloma cells. A myeloma is a plasma cell tumour and because the myeloma cells are transformed they will grow *in vitro* whereas B cells die in tissue culture. The product of the fusion of a B cell with a myeloma cell is called a **hybridoma**. Hybridomas combine the immortality of a myeloma cell with the antibody-producing capacity of the B cell. After fusion with the B cells the hybridomas can be selected in medium containing hypoxanthine, aminopterin and thymidine (HAT medium). The myeloma cells are deficient in the enzyme hypoxanthine guanine phosphoribosyl

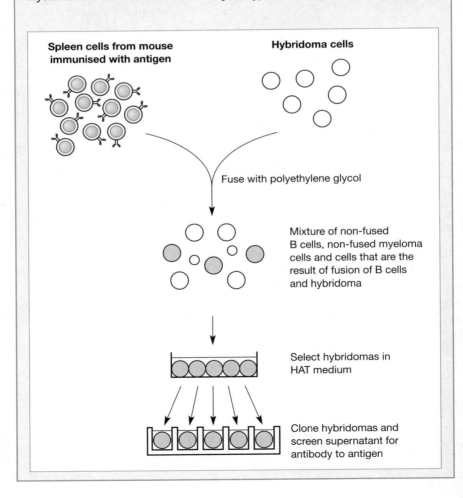

Spleen cells from mouse immunised with antigen

Hybridoma cells

Fuse with polyethylene glycol

Mixture of non-fused B cells, non-fused myeloma cells and cells that are the result of fusion of B cells and hybridoma

Select hybridomas in HAT medium

Clone hybridomas and screen supernatant for antibody to antigen transferase (HPRT) and cannot grow in the presence of aminopterin. The B cells are HPGRT+ but will die after a short time in culture. The hybridomas will be HPRGT+ because of the contribution by the B cell fusion partner and therefore will be the only cells to grow. Cloned hybridoma cells can be grown *in vitro* to produce large numbers of identical cells all producing the same antibody; for this reason the antibody produced by a hybridoma is called monoclonal antibody. It is now possible to produce large amounts of a particular monoclonal antibody, all of which is identical in both class and antigen specificity. This has revolutionised the production of antibody for research, technological and clinical use.

them into patients. Two variations have been considered. One is to use unmodified antibodies in the hope that they will bind to the tumour cells and recruit other components of the immune system to kill the tumour cells. This could be through complement fixation and lysis or the involvement of antibody-dependent cell-mediated cytotoxicity (Figure 15.6). A number of monoclonal antibodies have now been licensed for clinical use and more are in clinical trials (see Table 15.2).

A second approach with monoclonal antibodies is to attach them to toxic agents, which may be radioactive isotopes or toxins (Figure 15.6). These agents are too toxic to use directly but the theory is that the antibody will bind to antigens on the tumour cell and be internalised, leading to

Table 15.2 Antibodies being used as anti-tumour agents

Antibody	Target antigen	Tumour type	Status
Rituximab	CD20	Non-Hodgkin lymphoma	Licensed
Cetuximab	EGF-R	Colorectal	Licensed
Trastuzimab	Her2 (EGF-like receptor)	Breast cancer	Licensed
HuJ591	PSA	Prostate cancer	In clinical trials
Bevacizumab	VEGF	Colorectal (metastatic)	Licensed
^{131}I-tositumomab (Radio-Ab)	CD20	Non-Hodgkin lymphoma	Licensed
Gemtuzumab-Ozogamicin (Toxin-Ab)	CD33	Acute myeloid leukaemia	Licensed

EGF-R, epidermal-growth-factor receptor
PSA, prostate-specific membrane antigen
VEGF, vascular endothelial growth factor

Figure 15.6 Anti-tumour therapy with antibodies. Anti-tumour cell antibody can bind to the tumour cell and either fix complement, resulting in tumour cell lysis, or promote ADCC. Radioactive isotopes or toxins can be conjugated to the anti-tumour antibody, which then specifically targets the radiation or toxin to the tumour cell.

delivery of the toxic agent specifically to the tumour cells. Specific agents that have been considered for use include ^{131}I, bacterial toxins such as diphtheria toxin, and plant toxins including ricin, a very powerful toxin obtained from the castor bean. A number of these immunotoxins have been used in clinical trials but the overall results have been somewhat disappointing.

Tumour vaccines represent an approach to try to stimulate immune responses to tumours in the same way that microbial vaccines are used to stimulate responses against pathogens. Experimental evidence has shown that many tumours are antigenic; that is, they express antigens that can be the target of an immune response. There are, however, a number of features about tumours that enable them to avoid stimulating an immune response. Many tumours downregulate MHC molecules, enabling them to avoid recognition by T cells. Some tumours secrete immunosuppressive substances such as transforming growth factor-β. The aim of tumour vaccines is therefore to overcome the lack of immune responses to tumours and many approaches have been tried for this.

Peptide and DNA based vaccines

Many tumour antigens have been identified and in some cases peptides representing the T cell epitopes have also been characterised (see Table 15.3). Tumour antigens can be unique to individual tumours as a result of random mutations in the genes encoding the antigens or they can be shared tumour antigens. Shared tumour antigens are those that are expressed by tumours but not normal tissue or they are expressed in a quantitatively different way in tumour cells compared to normal. These have formed the basis of vaccines involving the injection of antigens or peptides with adjuvants (Figure 15.7), or the use of DNA vaccines. There are two conventional adjuvants

Table 15.3   Some examples of tumour antigens

Antigen	Type of antigen	Expression in tumour cells	Expression in normal cells
Antibody antigens			
Carcinoembryonic antigen (CEA)	Foetal	Colon + others	Yes
MUC-1	Aberrantly glycosylated mucin	Breast	Yes
p53	Mutated form of protein	Many	No
T-cell antigens			
MAGE-1, MAGE-3	Foetal	Melanoma, breast	Testis
ras	Mutated form of protein	Many	No
Tyrosinase	Differentiated cell product	Melanoma	Yes
Human papilloma virus E6 and E7 proteins	Viral gene product	Cervical carcinoma	Yes

Figure 15.7 Tumour vaccines. Known tumour antigens can be used as vaccines to try to stimulate immune responses against the antigen, which then destroy the tumour cells. The antigen/peptide is either injected *in vivo* to stimulate a response (top) or the patient's cells are incubated with antigen and injected back into the patient (bottom).

licensed for clinical use: alum and a squalene–oil–water emulsion. However, other agents, especially cytokines such as interleukin-2, granulocyte/macrophage colony-stimulating factor (GM-CSF) and interleukin-12, have been incorporated into tumour vaccines to try to boost the immune response. Many of these vaccines have been tested in clinical trials, especially for melanoma and breast cancer patients, using a number of antigens/peptides including the MAGE family and MUC-1.

DNA-based tumour vaccines use the same approach as with infectious agents (see Section 10.2.3) involving injection of DNA-encoding tumour antigens or peptides intra-muscularly or dermally via gene guns. It has been shown that the bacterial DNA acts as an adjuvant because the unmethylated CpG motifs are recognised by TLRs on dendritic cells and stimulate dendritic cells to become good stimulators of T cells.

Whole-cell based vaccines

One of the problems with the antigen/peptide approach is that for many tumours no tumour specific antigens have been identified and even some of the known tumour antigens are not expressed by tumours in all individuals. Another approach has been to use irradiated tumour cells or tumour cell lysates as vaccines. These have the advantage that both unique tumour and unknown shared antigens will have the potential to stimulate an immune response. Whole vaccines can be based on either cells derived from the tumours of the patient themselves or on tumour cell lines, although the latter, being derived from other individuals' tumours, will not stimulate responses against the patient's unique tumour antigens. Generally they are also given with an adjuvant. Whole vaccines have been tried in clinical trials for melanoma and pancreatic cancer, among others.

Transfected tumour cell vaccines

One of the findings from experimental and clinical trials with tumour cells as vaccines was that on their own they did not stimulate a very good immune response. An alternative approach to the use of adjuvants in whole vaccines is to transfect tumour cells. This approach has the advantage that it does not require knowledge of the particular tumour antigens but instead involves manipulation of the tumour cells so that they directly or indirectly stimulate immune responses. A typical approach is to take out some tumour cells and transfect them with genes that will make the tumour immunogenic or stimulate an inflammatory response (Figure 15.8). Tumour cells have been transfected with co-stimulatory molecules (CD80 or CD86) so that they can directly stimulate CD4 T cells. In experimental situations the transfected tumour cells stimulated a response that protected against both transfected and non-transfected tumours. The most successful experimental approach has been to transfect tumour cells with granulocyte macrophage colony-stimulating factor (GM-CSF). When injected back into animals the transfected tumour cells secrete GM-CSF, which attracts dendritic cells to the area. These dendritic cells process tumour antigens and carry them to draining lymph nodes, where they present antigen on their class II MHC, activated tumour-specific CD4 T cells and generate anti-tumour responses. These GM-CSF transfected tumour cells have also been shown to provide protection against both transfected and non-transfected tumours. It is vital that the transfected tumours also protect against non-transfected tumours because in the clinical situation you will be taking only a small proportion of the tumour cells for transfection and the transfected cells must be able to stimulate an anti-tumour response that is effective against non-transfected tumour cells.

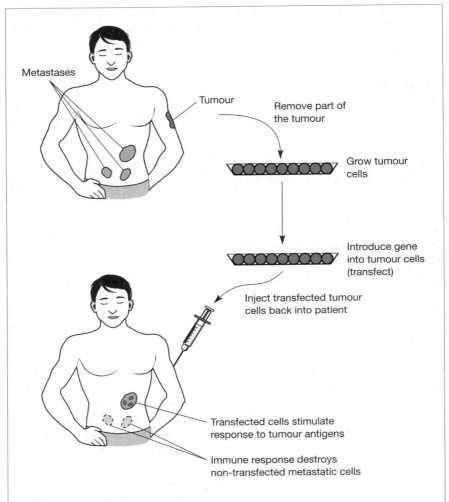

Figure 15.8 Transfected tumour cell vaccines. Part of the tumour is removed and grown in tissue culture. The tumour cells are transfected with a gene coding a particular protein that will make the tumour cells more immunogenic (e.g. GM-CSF or CD80). The transfected tumour cells are injected back into the patient in the hope that they will stimulate an immune response against the transfected cells that will also destroy non-transfected tumour cells.

Dendritic cell-based vaccines

One of the ways in which GM-CSF is thought to increase the immunogenicity of tumour cells is to activate dendritic cells so that they are better able to present tumour antigens and stimulate CD4 T cells. It is now possible to grow DCs in tissue culture by incubating peripheral blood monocytes with GM-CSF and IL-4. These dendritic cells can then be incubated with tumour cell lysates or tumour-specific antigens/peptides so that they take up antigen and present it on their class II MHC molecules. The dendritic cells are then injected back into the patients in the hope they will stimulate tumour-specific T cells. This approach has been tested in clinical trials for melanoma and gastro-intestinal cancers.

Cytokines, antigen-specific T cells and natural killer cells

Cytokines have been given systemically to try to boost immune responses, with limited success for some tumours. Although not strictly a vaccine, cytokines are being given with vaccines to increase immunogenecity (see above). Another use of cytokines, particularly IL-2, is to enable tumour-specific T cells to be grown *in vitro*. Lymphocytes extracted from tumours, so-called tumour-infiltrating lymphocytes (TILs), can be grown in tissue culture with IL-2 alone to expand the number of T-cells that have recently been activated with tumour antigen *in vivo*. Alternatively TILs or blood-derived T cells can be cultured with tumour antigen/peptide and IL-2 so that tumour antigen-specific T cells are stimulated by antigen and grow under the influence of IL-2. The cells can then be injected back into the patient to exert anti-tumour activity.

A final use of cytokines is in improving natural killer cell activity. Natural killer cells from blood can be grown in tissue culture and proliferate in the presence of high concentrations of IL-2 to become what are known as lymphokine-activated killer (LAK) cells. These LAK cells have enhanced tumour-killing activity and can be injected back into patients in the hope they will kill tumour cells.

It is evident from these examples that many innovative approaches are being used to induce or increase anti-tumour immune responses. Although they have been used in clinical trials, results have not been spectacular. Part of the problem is that the early clinical trials were in terminally ill patients with widespread untreatable tumours and the trials were not controlled. Results of the trials in terms of partial or complete tumour remissions were compared with historical remission rates for the particular tumours. However, results in some of the trials were encouraging enough for the approaches to be taken forward to larger controlled trials. This, combined with ever increasing knowledge of the interaction between the immune system and tumours, should lead to the development of effective tumour vaccines that can result in the elimination of established tumours.

15.4 Summary

- The most common form of transplant is an allograft, which is between two non-genetically related individuals of the same species. There is also considerable interest in xenografting tissue from other species to humans.

- Allografts stimulate strong immune responses against foreign MHC and non-MHC antigens. These can result in the generation of antibody, CD8 T cells and delayed-hypersensitivity responses against the graft.

- Graft rejection is prevented by immunosuppressive drugs, which are effective but can leave the recipient at increased risk of infection and the development of certain tumours.

- Immunotherapy of tumours involves two main approaches: the use of anti-tumour antibodies to destroy tumour cells and the development of tumour vaccines.

- Anti-tumour antibodies can be given in unmodified form or attached to radioactive or toxic agents. As yet, neither form has been very successful in clinical trials.

- Because tumours do not stimulate very good immune responses, two main types of tumour vaccine are being tested to try to stimulate anti-tumour immune responses. One type involves immunisation with identified tumour antigens to stimulate responses against the antigen. The other type involves transfecting tumour cells with genes that will make tumour cells stimulate an immune response that will be directed against both transfected and non-transfected tumour cells.

Index